Measuring behaviour

Measuring behaviour

AN INTRODUCTORY GUIDE

PAUL MARTIN
University Demonstrator in Animal Behaviour
Sub-Department of Animal Behaviour, University of Cambridge
and Fellow of Wolfson College, Cambridge

PATRICK BATESON, FRS
Professor of Ethology and Director of the Sub-Department of
Animal Behaviour, University of Cambridge
and Fellow of King's College, Cambridge

The right of the
University of Cambridge
to print and sell
all manner of books
was granted by
Henry VIII in 1534.
The University has printed
and published continuously
since 1584.

CAMBRIDGE UNIVERSITY PRESS
Cambridge
London   New York   New Rochelle
Melbourne   Sydney

Published by the Press Syndicate of the University of Cambridge
The Pitt Building, Trumpington Street, Cambridge CB2 1RP
32 East 57th Street, New York, NY 10022, USA
10 Stamford Road, Oakleigh, Melbourne 3166, Australia

First published 1986

Printed in Great Britain at the University Press, Cambridge

British Library cataloguing in publication data

Martin, Paul, 1958–
Measuring behaviour.

1.Animal behaviour – Measurement 2.Psychometrics
I.Title II.Bateson, Patrick
591.51   QL751.65.M4

0521323681

ISBN 32368 1 hard covers
ISBN 31184 5 paperback

1510572

CONTENTS

This book came about almost without our realising it. We needed something to give to our undergraduate students who were about to embark on projects that involved measuring behaviour. No single source was concise enough or easily accessible to them, so we prepared some notes. These grew longer and longer, they generated more and more comments and, before long, we found ourselves doing something that neither of us had planned.

As the shape of the book started to emerge, we felt that its recommendations ought to represent, as far as possible, the collective view of the practitioners of the subject, rather than merely our own opinions. Therefore, a draft version was read by a large number of friends and colleagues. The eventual book was immeasurably improved by their advice, comments, criticisms and suggestions. Inevitably, some issues still provoke substantial disagreement; we have attempted to indicate where this is the case and where we have departed from common practice.

Some of our reviewers asked us to put in detailed examples, to illustrate the general points and to show in particular cases how methods are selected and used. We decided not to do so to the extent they would have liked for several reasons. First, we wanted to keep the book as concise as possible, in order to make the information in it readily accessible. Second, we wished to draw attention to general principles of measurement and analysis. Many of the methods and ideas are relevant to a wide variety of disciplines, involving field and laboratory studies of both humans and non-human species. We were worried that specific examples might indicate to potential users interested in other problems or species that the methods were unsuitable for their purposes. Third, we intended the book to be used as a guide and consulted as the need arises throughout a research project. The best examples are then provided by the user's own work.

In any event, we are deeply grateful to our friends and colleagues who have so willingly shared their knowledge and given so much thought and time in helping us. We warmly thank the following for reading and commenting on an earlier version of this book: Stuart Altmann, Bob Baldwin, Marc Bekoff, Donald Broom, Gordon Burghardt, Tim Caro, Neil Chalmers, Tim Clutton-Brock, Patrick Colgan, Eberhard Curio, Nick Davies, Amanda Dennis, Robin Dunbar, Judy Dunn, Stephen Emlen, Tim Halliday, Sandy Harcourt, Robert Hinde, Felicity Huntingford, Phyllis Lee, Brian McCabe, Cathie Marsh, Nicholas Owens, Michael Reiss, Thelma Rowell, John Sants, Riccardo Scartezzini, Peter Slater, Peter Smith, Joan Stevenson-Hinde, Kelly Stewart, Carel Ten Cate and Dennis Unwin. We are also grateful to Les Barden for drawing the figures.

Finally, we thank Robin Pellew who has guided us so well, both in his capacity as a scientist and as biology editor at Cambridge University Press, throughout the production of this book.

We should welcome the advice and criticism of anybody who uses this book, to help us improve future editions. Please send comments to us at the following address: Cambridge University Sub-Department of Animal Behaviour, Madingley, Cambridge CB3 8AA, UK.

Cambridge, September 1985   Paul Martin
                                                 Patrick Bateson

1 *Formulate an initial question and make preliminary observations.* The question(s) asked may stem from previous knowledge or observations and may initially be rather broad or quite specific. A period of preliminary observations provides an essential opportunity to become familiar with the subjects and their behaviour, and to develop hypotheses and methods of measurement (Chapters 1, 2 and 3).

2 *Formulate hypotheses and make predictions.* On the basis of preliminary observations and existing knowledge, formulate hypotheses: the more the better. These should make specific predictions. Choose observations or experiments that will best distinguish between the competing hypotheses. Formulating hypotheses is a creative process requiring imagination (Chapters 1 and 2).

3 *Choose behavioural measures and a research design.* Be reasonably selective. Choose measures that are relevant to the question and which will help distinguish between different hypotheses (Chapters 2 and 3). Hypotheses can be tested by observing natural variation as well as by performing experiments.

4 *Define each measure.* Definitions should be clear, comprehensive and unambiguous. Write down the definitions before starting to collect data (Section 3.D).

5 ***Select the appropriate recording methods.*** These will depend on the nature of the behaviour to be recorded (Chapters 4 and 5). In practice, steps 3, 4 and 5 are inter-related, since the recording methods available often influence the choice and definition of categories.

6 ***Practise the recording methods.*** Assess the reliability and validity of each category. Drop categories that are clearly unreliable and irrelevant. Measure inter- and intra-observer reliability at the beginning and end of data collection. Be prepared to add new categories and to re-define categories in the light of preliminary observations and pilot measurements (Chapter 6).

7 ***Collect data.*** Use the same measurement procedures throughout. Don't collect more data than are needed to answer the question.

8 ***Analyse the data.*** Carry out exploratory analysis (for example, by drawing graphs and calculating summary statistics) before confirmatory analysis. Don't sacrifice clarity for complexity by using unnecessarily complicated statistics. Don't draw more conclusions than the data actually support. Distinguish between generating new hypotheses from the data and testing them. Present the results clearly, in summary form (Chapter 9).

1

General issues

A Why measure behaviour?

In addition to its intrinsic interest, the study of behaviour is both intellectually challenging and practically important. Animals use their freedom to move as one of the most important ways in which they adapt themselves to the conditions in which they live. These adaptations take many different forms such as finding food, avoiding being eaten, finding a suitable place to live, finding a mate and caring for young. Each species tends to have special requirements and often the same problem is solved in different ways by different species.

Even though much is already known about such adaptations and the ways in which they are refined as individuals gather experience, a great deal remains to be discovered about the diversity and functions of behaviour. The principles involved in the evolution of increasingly complex behaviour and the role that behaviour itself has played in shaping the direction of evolution are still not well understood. Understanding how behaviour patterns have arisen and what they are for will only come from the comparative study of different species and by relating behaviour to the social and ecological conditions in which an animal lives.

What about mechanism? Molecular and cellular approaches to biology have made remarkable progress in the last 30 years. The neurosciences are beginning to uncover how nervous systems work and the long-standing goal of understanding behaviour in terms of underlying processes is becoming attainable at last. So why bother with the measurement of *behaviour* (by which we mean the actions and reactions of whole organisms)? The answer is illustrated by a simple analogy. Perfect knowledge of how often each of the letters on this page occurs would give no indication of their meaning until the letters are formed into words and the words are formed into sentences. Each successive level of organisation has properties that cannot be predicted from knowing the lower levels of organisation. Even when the understanding of the underlying neural elements is complete, it will not be possible to predict how they perform as a whole without first understanding what they do as a whole – and that means knowing how the whole organism behaves. Understanding the mechanisms underlying behaviour requires first understanding the behaviour itself. For all these different reasons we believe that thorough description and analysis of behaviour is going to remain an essential part of biology for a long time to come.

B The scope of this book

Sometimes it is possible to carry out behavioural research simply by relying on written descriptions of what the subjects do. Usually, though, worthwhile research will require that at least some aspects of the behaviour are measured. (By *measure* we mean quantify by assigning numbers to observations according to specified rules.) Therefore, measurement of behaviour, both in the laboratory and in the field, is required by virtually all behavioural biologists and psychologists. In this book we are primarily concerned with the methods based on *direct observation* of behaviour which were developed for recording the activities of non-human species. These methods are not only applicable in advanced academic research. They may also be readily used in the behavioural projects that are commonly offered in university teaching courses and, increasingly, in schools. Moreover, the techniques may fruitfully be applied in studies of human behaviour – even though they do not deal with some important issues, such as the measurement and analysis of language – and therefore have important uses in the social and medical sciences.

This book is intended as a guide to all those who are about to start work that involves the measurement of directly observed behaviour. We hope it will also be useful for those who seek to refresh their memories about both the possibilities and the shortcomings of available techniques.

Those who have never attempted to measure behaviour may suppose from the safety of an armchair that the job is an easy and straightforward one, requiring no special knowledge or skills. Is it not simply a matter of writing down what happens? In sharp contrast, those attempting to make systematic measurements of behaviour for the first time are often appalled by the apparent difficulty of the job facing them. How will they ever notice, let alone record accurately and systematically, all that is happening? The truth is that measuring behaviour *is* a skill, but not one that is especially difficult to master, given some basic knowledge and an awareness of the possible pitfalls. The purpose of this book is to provide the basic knowledge in a succinct and easily understood form, enabling the beginner to start measuring behaviour accurately and reliably. A great deal of high quality behavioural research can be done without the need for specialised skills or elaborate and expensive equipment.

Inevitably in such a short book, we have dealt with many complex and contentious issues rather briefly, and in some cases the advice we offer is based on opinions that are not universally shared. We hope that readers will note our cautions and, where necessary, explore the issues in greater depth than is possible here. To help with this, we have suggested some further reading at the end of each chapter and have provided annotated references at the end of the book. We also suggest that, wherever possible, beginners discuss their proposed methods with an experienced observer and seek advice before proceeding too far in their studies.

C   Different approaches to studying behaviour

Not all scientists study behaviour in the same way or ask the same sorts of questions. Historically, **psychology** (which originally grew out of the study of the human mind) was separated from **etology** (the biological study of behaviour) in terms of the methods, interests and the origins of the two sciences. In this century, comparative and experimental psychologists have tended to focus mainly on questions about the proximate causation of behaviour (so-called 'how' questions), studying general processes of behaviour (notably learning) in a few species under laboratory conditions. In contrast, ethologists have their roots in biology and have asked questions not only about how behaviour is controlled but also about what behaviour is for and how it evolved ('why' questions).

Biologists are trained to compare and contrast diverse species. Impregnated as their thinking is with the Darwinian theory of evolution, they repeatedly speculate on the adaptive significance of the differences between and within species. Indeed, many ethologists are primarily interested in the biological functions of behaviour and are wary of proceeding far with laboratory experiments without first understanding the function of the behaviour in its natural context. Studies in unconstrained conditions of animals, and increasingly of humans, have been an important feature of ethology and have played a major role in developing the distinctive and powerful methods for observing and measuring behaviour. In contrast, psychologists have traditionally placed greater emphasis on experimental design and quantitative methods. Even so, it would be a mistake to represent modern ethology as non-experimental. A great many people who call themselves ethologists have devoted much of their professional lives to laboratory studies of the control and development of behaviour.

Field studies which related behaviour patterns to the social and ecological conditions in which they normally occur led to the development of a new subject, **behavioural ecology**. Another sub-discipline, **sociobiology**, brought to the study of behaviour important concepts and methods from population biology and stimulated further interest in field studies of animal behaviour. All these sub-disciplines are showing signs of merging and, for simplicity, we shall refer henceforward to ethology, behavioural ecology and sociobiology collectively as **behavioural biology**.

Modern behavioural biology abuts many different disciplines and defies simple definition in terms of a common problem or shared ideas. The methods developed for the measurement of behaviour are used by neurobiologists, behaviour geneticists, social and developmental psychologists, anthropologists and psychiatrists, among many others. Considerable transfer of ideas and a convergence of thinking has occurred between behavioural biology and psychology, from which both subjects have greatly benefited. For example, experimental methods developed by psychologists are being used by behavioural ecologists interested in how animals forage for food. Conversely, many of the observational methods developed by behavioural biologists have proved highly effective in studying the behaviour of children.

D The four problems

It is important to realise that a number of quite different types of question can be asked about behaviour. Probably the most useful classification was formulated by the ethologist Niko Tinbergen, who pointed out that four distinct problems are raised by the study of behaviour:

1 **Proximate causation** or **Control** ('How does it work?'). How do internal and external causal factors elicit and control behaviour in the short term? For example, which stimuli elicit the behaviour pattern and what are the underlying neurobiological, psychological or hormonal mechanisms regulating the animal's behaviour?

2 **Development** or **Ontogeny**. How did the behaviour arise during the lifetime of the individual; that is, how is behaviour assembled? What internal and external factors influence the way in which behaviour develops during the lifetime of the individual and how do the developmental processes work? How do an individual's genes and experience interact during the assembly of its behaviour?

3 **Function** ('What is it for?'). What is the current use or survival value of the behaviour? How does behaving in a particular way help the individual to survive and reproduce in its physical and social environment?

4 **Evolution** or **Phylogeny**. How did the behaviour evolve during the history of the species? What factors might have been involved in moulding the behaviour over the course of evolutionary history? Note that evolutionary questions are concerned with the historical origins of behaviour patterns, whereas functional questions concern their current utility. The two are frequently muddled. Questions of function and evolution are sometimes referred to as 'ultimate' questions when contrasted with proximate causation.

The Four Problems can perhaps best be illustrated with a commonplace example. Suppose we ask why it is that drivers stop their cars at red traffic lights. One answer would be that a specific visual stimulus (the red light) is perceived, processed in the central nervous system and reliably elicits a specific response (changing gear, applying the brake and so on). This would be an explanation in terms of proximate causation. A different answer is that individual drivers learn this rule from books, television and driving instructors. This is an explanation in terms of ontogeny. A functional explanation is that drivers who do not stop at red traffic lights are liable to have an accident or, at least, be stopped by the police. Finally, an 'evolutionary' explanation would deal with the historical process whereby a red light came to be used in many countries as a way of stopping traffic at road junctions.

Tinbergen's Four Problems are logically distinct and should not be confused with one another. Nonetheless, it can be helpful to ask more than one type of question at the same time. For instance, asking what a particular behaviour pattern is for (its function) can sometimes help to suggest ways of studying the underlying mechanism (proximate causation).

E Choosing the right level of analysis

The form of measurement used should depend on the nature of the problem and the questions posed. Conversely, the sorts of phenomena that are uncovered by a study will inevitably reflect the methods used. To a lesser extent, the choice of behavioural categories and recording methods should also take into account the statistical methods that will eventually be used to analyse the results. Sometimes results are painstakingly collected, only to be found unsuitable in form for the statistical tests required to analyse them. Consulting a statistician *before* starting to collect data is generally a wise precaution. If possible, seek advice from a statistician who is familiar with behavioural research.

The pattern of events an observer notices is affected by spatial scale: being close to a subject reveals details that would not be seen at a distance. Similar arguments apply to the time scale of measurements. For example, the value of 'freezing' time was demonstrated when the first photographs of galloping horses were taken in the nineteenth century. Hitherto, artists had conventionally

Fig. 1.1. Two representations of how a horse gallops. Until photography provided a way of 'freezing' time, the upper picture was believed to be the correct one.

WRONG

RIGHT

painted galloping horses as having their front and back legs extended simultaneously. Photographs revealed that, in fact, a horse never does this (Fig. 1.1).

A detailed analysis is only appropriate for answering some questions, and a full understanding does not necessarily emerge from describing and analysing behaviour in minute detail. While a microscope is an invaluable tool in some circumstances, it would be useless for reading a novel. In other words, the cost of gaining detail can be that higher-level patterns, which may be the most important feature, are lost from view. For example, recording the precise three-dimensional pattern of movements for each limb may be desirable for certain purposes, but higher-level categories such as 'walking' or 'running' are usually more appropriate. On a more mundane level, recording large amounts of unnecessary detail may obscure broader issues simply by presenting the researcher with overwhelming quantities of data to analyse and interpret.

F Choosing the right species

For many people, choosing which species to study is not an issue. For example, they are most interested in and only want to study humans; they study an animal because only a handful of that particular species is left on the planet; they have no option other than to work on the species that is bred in a particular laboratory; and so forth. Nonetheless, when choice is possible it is worth giving considerable thought to the advantages and disadvantages of the vast number of species available. The wealth and diversity of zoological material is so great that time invested in finding a species that is suitable for the problem to be investigated is likely to be amply repaid later in the study.

What follows is a list of points to think about, but each problem has different requirements and only some of the points may be relevant in any one case.

(*a*) Is it easily seen in its natural habitat or readily available for study in captive conditions? If it has to be imported, are the conditions for collecting it in the country of origin ethical? Will long delays ensue because of quarantine requirements?

(*b*) Is it tolerant of human presence? Does it handle well if it is to be kept in captivity or is hand-reared? Does it breed well in captivity? If it is to be kept in captivity, does it have any special dietary requirements? Is it small enough to make feeding and housing financially feasible if large numbers are required? Are there remedies for the diseases it is likely to get?

(*c*) What are its life-history characteristics such as gestation period, age of independence and age of sexual maturity? Is its life-span long enough to make repeated measurements possible but short enough to make studies of development practicable?

(*d*) Is much else known about its natural history, anatomy, and physiology? Is there an extensive biological and behavioural literature available for this species?

(*e*) Is anything known about its genetics? Would controlled pro-grammes of breeding be possible?

(*f*) Is much known about its behaviour? At what time of day is it typically active? (A surprising number of people study nocturnal animals when they are least active.) How solitary or gregarious is it?

(*g*) Is its behaviour such that it will be suitable for the particular problem which is to be investigated? Does it move slowly enough to make observation relatively easy yet fast enough to make observation rewarding?

(*h*) If a general problem relevant to many species is to be studied, is the particular species chosen a suitable model? If the species is to be studied for its own sake, can knowledge of it contribute to an understanding of human behaviour?

(*i*) Are there opportunities for comparing its behaviour with that of other closely related species?

Sources of information. Unless expert help is sought, many of these questions cannot be answered without spending time in a library. Some handbooks review the knowledge of certain taxonomic groups species by species; for example, Eisenberg (1981) for mammals and Cramp & Simmons (1977–1985) for birds. Most of the relevant abstracting journals (for example, *Animal Behaviour Abstracts*,

Biological Abstracts and *Zoological Record*) index their contents by species as well as under other headings. The *Biology Data Book* (Altman & Dittmer, 1972–1974) contains useful tables of comparative information and *The UFAW Handbook* (UFAW, 1976) provides invaluable guidance on the care of animals in captivity.

G Anthropomorphism

Humans readily interpret the behaviour of other species in terms of their own thoughts and intentions, and observing animals often leaves the impression that the animals know what they are doing. However, subsequent analysis frequently reveals that seemingly complex and purposive behaviour can be produced by simple mechanisms that do not involve conscious thought, motives or intentions. For example, a woodlouse will move about briskly when it is in a dry environment, and sluggishly (or not at all) when in a humid environment. The animal appears to seek out damp places in a purposeful manner, but its response can be explained in terms that are no more complicated than those of an electric heater controlled by a thermostat.

Using human thoughts and intentions as *explanations* for animals' actions can impede further attempts to understand the behaviour. In general, therefore, it is wise to start by obeying the injunction to explain behaviour in the simplest possible way until there is good reason to think otherwise (**Lloyd Morgan's Canon**).

Nonetheless, slavish obedience to a maxim tends to sterilise imagination and although the possibility of anthropomorphism must be acknowledged, an over-emphasis on its dangers can constrain research. A scientist who never thinks of an animal as though it were a human is liable to miss much of the richness and complexity of its behaviour. If an animal is invariably thought of as a piece of clockwork machinery then some of its most interesting attributes may be overlooked (see, for example, Griffin, 1984).

The general suggestion, then, is to muster every possible type of mental aid when generating ideas and hypotheses, but to use the full rigour of analytical thought when testing them.

Another fundamental point is that other species occupy different **perceptual worlds** from humans – in other words, their sensory abilities may be radically different from our own. For example, many rodents communicate using ultrasonic vocalisations, some insects can detect ultra-violet light, some snakes can detect prey using infra-red sensors and many species have highly developed powers of olfaction. Humans occupy a perceptual world that is dominated by colour vision, but this is not true for all other species. Thus, an animal may be oblivious to a visual stimulus that seems obvious to a human or, conversely, may respond to a stimulus that a human cannot detect.

H **The steps involved in studying behaviour**
Studying behaviour involves a number of inter-related processes:

1 *Ask a question*. At the very outset of a study, it is important to have a clear idea of the general issues to be investigated. Before any scientific problem can be studied, some sort of question must be formulated. The question may initially be a broad one, stemming from simple curiosity about a species or a general class of behaviour, such as 'What is the sexual behaviour of this species like?'. The enormous value of broad description arising from sheer curiosity certainly should not be underestimated. Alternatively, it may be possible at an early stage to formulate a much more specific question based on existing knowledge and theory, such as 'Do big males of this species acquire more mates than small males?'. Not surprisingly, research questions tend to become more specific as more is discovered about a particular issue.

The particular choice of question (or questions) may be influenced by a variety of factors, including the investigator's previous knowledge, interests and observations made in the course of other research. Quite often, the impetus for a study stems from little more than a hunch.

2 *Make preliminary observations and formulate hypotheses.* We shall discuss these two processes together, since in practice they are often inseparable and concurrent. Hypotheses are essentially specific questions. Formulating hypotheses is a creative process, requiring imagination and some knowledge of the issues involved. In some cases, a considerable amount of descriptive information about a problem may be required before useful and interesting hypotheses can be formulated, while other problems lend themselves more easily to setting out specific hypotheses at an early stage in their investigation.

It is not possible to give specific advice on how to formulate good hypotheses, any more than advice can be given on how to write good literature or paint good pictures. In general, the more competing hypotheses that are formulated the better. The danger with having only one hypothesis is that it may be difficult to reject.

A period of unfettered preliminary observation is often of great importance in formulating interesting hypotheses and should be regarded as an essential part of any study. Jumping straight in and collecting 'hard data' from the very beginning is not usually the best way to proceed.

3 *Make predictions from the hypotheses.* A clear hypothesis should, by a process of logical reasoning, give rise to one or more specific predictions that can be tested empirically. The more specific the predictions are, the easier it usually is to distinguish empirically between competing hypotheses and thereby reduce the number of different ways in which the results could be explained.

4 *Identify which behavioural variables need to be measured in order to test these predictions.* The form of the research and the variables that are measured should be chosen so as to provide the best test of the predictions and to allow hypotheses to be rejected.

5 *Choose suitable recording methods for measuring these behavioural variables.* No observer can record behaviour without selecting some features from the stream of events and ignoring others. This selection inevitably reflects the observer's own interests, preconceptions and hypotheses. It simply is not possible to record everything that happens, because any stream of behaviour could, in principle, be described in an enormous number of different ways. The choice of which particular aspects to measure, and the way in which this is done, should reflect explicit questions.

Making the transition from thinking about a problem and formulating hypotheses to tackling it empirically is found by many people to be one of the most difficult parts of research.

6 *Collect sufficient data.* Stop collecting data when there are sufficient to provide clear answers. Once they have started collecting data in a systematic way, many people find it difficult to stop.

7 *Employ the appropriate statistical tools, both for presenting and exploring the data, and for testing the hypotheses.* Carry out exploratory data analysis to obtain the maximum amount of information from the data and to discover unexpected results that generate new questions. Use confirmatory analysis to test hypotheses. Distinguish between testing existing hypotheses and generating new ones. Do not draw more conclusions than the data support, but do try to formulate a list of questions and ideas suggested by the data which could form the basis of future research.

The various processes involved in measuring behaviour are represented in Fig. 1.2.

A major aim of scientific research is to help distinguish between competing hypotheses and thereby reduce the number of different ways in which the natural world can be accounted for. In this respect, there is no fundamental distinction between *experimental* research (in which variables are actively manipulated) and purely *observational* research, since both can generate empirical data that distinguish between competing hypotheses. Some behavioural research is purely observational, but this does not make it any less scientific. Remember that in some areas of science, such as astronomy and geology, conventional experiments are rarely (if ever) possible, yet detailed quantitative hypotheses are regularly formulated and tested by observation.

Measuring behaviour in order to answer one set of questions will inevitably produce results that in turn generate new questions. In this sense, scientific research has a cyclical nature (see Fig. 1.3).

The successful scientist is likely to be one who can combine a purposeful approach to tackling an initial set of questions with the ability to recognise and respond opportunistically to new questions that arise during the course of research. A study is unlikely to be fruitful if it remains completely open-ended and never focuses on

Fig. 1.2. The processes involved in studying behaviour.

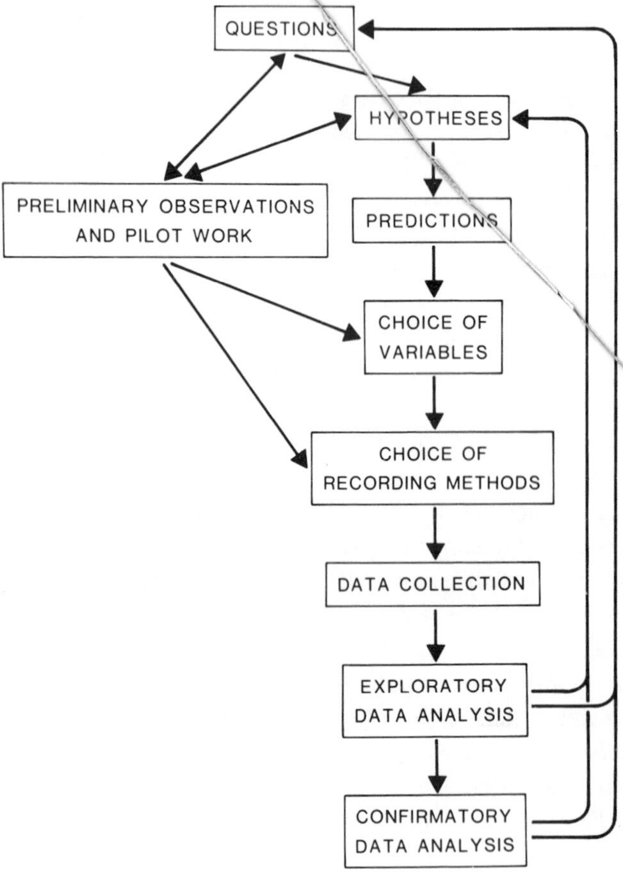

any specific issues. Conversely, if one problem is pursued in a rigid and inflexible way to the exclusion of all else then potentially important new ideas and observations may be missed.

I Further reading

Medawar's (1981) short book about science and scientists is a delight to read and is packed with common sense and wit. The relations between ethology and psychology are discussed by Jaynes (1969), Hutt & Hutt (1970), Lockard (1971), Slater (1980) and Hinde (1982). The Four Problems and other central principles of ethological thinking are to be found in the seminal paper by Tinbergen (1963). See also Hinde (1970, ch. 2; 1982, ch. 1) and Hutt & Hutt (1970, ch. 2). Beach's (1950) famous essay provided a powerful rebuttal to an earlier generation of psychologists who saw the white rat as a 'general purpose' experimental subject, equally suitable for studying all aspects of behaviour. For a defence of the 'natural history' approach to biology – as opposed to the more common approach of picking a species because it is suitable for studying a general problem – see Gans (1978).

Fig. 1.3. The cyclical nature of scientific research.

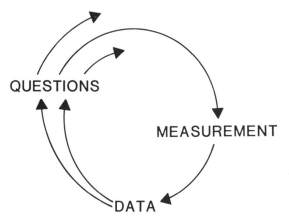

2

Research design

A Rules for research

Formal prescriptions about the way in which scientific research should be conducted often fail to capture the flair, intuition and idiosyncrasies of the best scientists. Therefore, advice on how to organise thought and carry out research has to be given and taken with caution. Some people start studying a particular species simply because they find the animal interesting. Other scientists are initially more interested in a theoretical problem and, if they are wise, choose an animal that is especially suitable for studying that problem. It is not obvious that one approach is better than the other and, indeed, they are usually complementary.

As we pointed out in Chapter 1, empirical data that distinguish between competing hypotheses can be obtained by observing natural variation as well as by performing controlled experiments. Thus, for many purposes, it is unnecessary to draw a rigid distinction between observational and experimental research. Many questions about behaviour are most appropriately answered by observational research. Moreover, in order to be effective, experimental research almost invariably needs to be preceded by observation. Knowledge of the normal behaviour of healthy animals, preferably in their natural environment, is an essential precursor to experimental research (see Chapter 7 on field studies). One of the distinguishing features of the ethological approach to studying behaviour has been its emphasis on combining observation with experiment (see 7.D).

B Effects of the observer on the subject

The observer is rarely invisible and may have a profound effect on the subjects, in both laboratory and field studies. Even subjects that neither react with alarm to the observer's presence nor attempt to escape from the observer may, nonetheless, alter their behaviour in subtle ways. A strong conviction that the subjects are not affected by the observer's presence can often turn out to be mistaken.

In field studies, disruption can be reduced by using hides or blinds to conceal the observer. If the observer cannot approach the hide without being noticed by the subjects, an accomplice may have to place the observer there and walk ostentatiously away – although some animals may not be taken in by such a ruse. Of course, restricting the observer to a hide, even a mobile one such as a vehicle, may mean that some of the most interesting aspects of an animal's behaviour are missed. Consequently, in many studies observers spend long periods simply accustoming the subjects to their presence, a stratagem that generally seems to work well. Nonetheless, the impression that well-habituated subjects are not affected by the observer's presence is difficult to verify and should be treated with some scepticism. This point is especially relevant to field studies of behaviour (see Chapter 7).

The observer's presence may introduce subtle bias even though the subjects appear to be well habituated. For example, some activities (such as play or sexual behaviour), or some individuals (such as juveniles) may be more affected than others by the observer's presence. Similarly, even though the subjects may be habituated to the observer's presence, their predators or prey may not.

In the laboratory, a one-way screen made of half-silvered glass, dark-tinted Perspex or muslin can be placed between the observer and the subject. This technique for concealing the observer relies on the illumination being much brighter on the subject's side of the screen. Another method is to place an angled mirror above the subject and, on the assumption that it does not look up, watch from a position where direct visual contact is impossible. Of course, even if the observer cannot be seen, the subject may still be able to hear or smell the observer. A third possibility, which solves this problem, is to use a concealed television camera and watch the behaviour on a remote screen. The ability of closed-circuit television to capture detail is rarely as good as direct observation, but this method has the advantage that a permanent record can be kept on videotape.

In laboratory studies it is generally a good idea, where possible, to test animals in their home cages. Transferring an animal from its home cage to a strange environment in order to observe it may disrupt its behaviour considerably. If testing or observing in the home cage is impossible, it may be necessary to ensure that subjects are well habituated to the test situation before data are collected.

Psychologists and sociologists have known for many decades that changes in the behaviour of their subjects sometimes result not from the effects of any experimental manipulation, but merely from the attention paid to them by the experimenter. In psychology and sociology this source of error is known as the **Hawthorne Effect** (see Sprinthall, 1982, p. 223).

C Experimental design: controls, experimenter bias, replication and order effects

1 *Controls.* The point of an experiment is to find out whether varying one condition causes a particular outcome, thereby reducing the number of plausible alternative hypotheses that could be used to account for the results. The aim of the simplest experiment, then, is to vary just one condition (the **independent variable**) and measure the effect on one or more outcome measures (or **dependent variables**), whilst holding all other conditions constant. (In more complex experiments two or more conditions might be varied simultaneously and many dependent variables measured.) The effects of varying the condition (the so-called **treatment effect**) are measured for one group of subjects (the experimental or treatment group) and are compared with those for a control group of subjects which do not receive the experimental manipulation but are in all other respects similar.

For example, suppose we wished to measure how the hormone testosterone affects the sexual behaviour of male rats. The experimental subjects would each receive an injection containing a certain dose of testosterone. The outcome measures (or dependent variables) would be one or more measures of sexual behaviour. Now, any effect an injection has on sexual behaviour could be caused by the hormone, but it could also be explained by other **confounding factors** such as the effects of handling by the experimenter, the discomfort of the injection or the effects of the substance in which the hormone is dissolved (the so-called vehicle). Simply giving the control animals no injection would thus not constitute a good control, since these other confounding factors would also be different for the two groups. Therefore, the control group should also receive an injection, but of the vehicle substance only.

In behavioural experiments it is usually difficult to vary one condition without varying something else as well. Part of the art of good experimenting lies in (*a*) picking the appropriate control groups and (*b*) randomising the effects of confounding variables. In these ways it becomes possible to distinguish between the effects of variables that would otherwise be confounded with each other.

Suppose an experiment is designed to test the effects of varying one condition (X), but when X is varied another condition (Z) also varies. Any observed effect of varying X could in fact be due to variations in X or in the confounding variable Z. Now, it may be difficult to vary X without varying Z, but the converse need not be true. If so, the obvious control condition would be to vary Z without varying X. For example, Vauclair & Bateson (1975) investigated whether experience of pecking at seeds (X) enabled newly hatched domestic chicks to peck more accurately. However, any improvement in pecking accuracy might also be due to the non-specific effects of exposure to patterned light (the confounding variable, Z). Experience of pecking (X) is confounded with exposure to patterned light (Z), but the converse is not necessarily true. Therefore, the effects of pecking and exposure to light were disentangled by exposing a control group to patterned light whilst preventing them from pecking.

Individuals can differ enormously in their behaviour and it is often desirable to remove the confounding effects of the constellation of factors that distinguish one individual from another. The simplest and most effective way of doing this is to use the individual as its own control, by testing the same individual under both the experimental and control conditions on separate occasions. This constitutes a **matched-pairs** design, in which the scores are the *differences* between the experimental and control conditions for each subject. One problem with this design is that the experimental treatment may have a lasting effect, so that subsequent measurements do not reflect a true control condition. In addition, the order in which the experimental and control conditions are administered must be randomly determined, so that roughly equal numbers of subjects will receive the experimental and control conditions first (see section 4, below, on order effects).

2 *Experimenter bias.* Scientists often have strong expectations about the outcome of an experiment, even if they are not consciously aware of them. This source of bias can be controlled for by ensuring that the person making the measurements is unaware of which treatment each subject has received until after the experiment is over. This procedure is referred to as running a **blind** experiment. If the experimenter is not blind to the treatment when measurements are made, two sorts of bias can arise. First, the experimenter may unconsciously provide clues to the subjects which influence their behaviour in a particular way. The most famous example of this was the case of **Clever Hans**, a performing horse that seemed to be able to count. Only as a result of careful experiments under blind conditions was it found that the horse was not able to count but was, in fact, responding to subtle cues unconsciously given by its trainer.

In addition to directly affecting the subject's behaviour, experimenter bias can also arise when recording or analysing the results. For example, several experiments have apparently demonstrated that play improves children's abilities to solve problems or to think creatively. However, a careful review of these experiments has pointed to some serious methodological problems (Smith & Simon, 1984). In many of the experiments the same person administered the different treatment conditions, tested the children and scored their performance, allowing experimenter bias to arise at all three stages. Clearly, an experimenter who was aware of the hypothesis, and believed that play would enhance the children's performance, might administer the conditions or the tests differently and might show unconscious bias in scoring the results. In this type of experiment, as in much behavioural research, measuring behaviour can involve a considerable amount of interpretation and judgement by the observer. More recent experiments, which have controlled for experimenter bias, have found little evidence that play is superior to other forms of experience in improving children's problem-solving or creative thinking abilities (e.g., Simon & Smith, 1983).

In summary, then, unless experimenters are unaware of the treatment received by each subject, they may unintentionally affect the subjects' behaviour and bias scores in the expected direction. The cumulative effect of many such minor biases may be an apparent significant difference between the experimental and control groups. These effects are often much larger than many people realise (Rosenthal & Rubin, 1978) and the only sure way to minimise them is for the person making the measurements to be unaware of how the subjects have been treated.

If the subjects of a study are humans, they too may introduce bias into the results if they are aware of the group they are in or the treatment they have received. If at all possible (and ethically acceptable), human subjects should not be aware of which group they are in until after the study is over. An experiment in which neither the person making the measurements nor the subjects know the treatment each subject has received is called a **double blind** experiment. This type of experimental design is widely used, for example, in assessing the clinical effects of drugs.

3 *Replication.* Incorrect theories present no great threat to science because they will soon be shown to be at odds with reality. However, incorrect empirical findings can be widely accepted as 'fact' unless attempts are made to replicate the study that reported them. Where empirical results bear crucially on an important theory, it is common for measurements to be replicated. However, the immense diversity, variability and complexity of behaviour, together with the relative dearth of quantitative theories in some areas of the behavioural sciences, means that the pressure to replicate behavioural studies is often lacking. Some of the most cherished empirical findings, widely cited in textbooks of animal behaviour, have not been replicated.

Replication can occur at two levels. First, an attempt can be made to duplicate a particular study or experiment exactly, using the same species and precisely the same measures, procedures and conditions, in order to ensure that the original data were correct. This is referred to as **literal replication** (Lykken, 1968). One problem with behavioural research, however, is that no two samples of individuals are identical: even the same sample measured under the same conditions may vary with time or with experience. Furthermore, different individuals and different strains of the same species may differ considerably in their behaviour. Thus exact replication under truly identical conditions is rarely feasible.

A second approach to replication, known as **constructive replication**, is to produce convergent results that lend support to the initial study's findings, but using different measurement procedures – perhaps different measures, experimental conditions or even a different species. In other words, no explicit attempt is made to duplicate the original study in terms of measurement procedures; the emphasis is instead on whether the conclusions are supported empirically.

The results of any scientific study must always be reported with complete honesty.

4 *Order effects.* Repeated treatment or testing of the same subject may markedly influence its behaviour. For instance, 3-day-old domestic chicks initially avoid novel moving objects. With successive presentations, however, the amount of time spent avoiding the object first increases and then declines, until eventually the object is ignored or even followed (Bateson, 1964). Many processes such as arousal, sensitisation, conditioning, fatigue and habituation may contribute and interact so that the changes in the subject's responsiveness over time are not simple. These changes may themselves be of interest. But if they are ignored, they can mean that apparently identical experimental conditions do not have the same effects on the subject, so that measurements made in sequence are not comparable. The point to remember is that once an individual is tested, it becomes a somewhat different individual.

Order effects can sometimes be balanced between different groups, some individuals getting one order of presentation while others get another, so that every possible order is used. Alternatively, the order can be randomised, using a Latin square design (e.g., Lehner, 1979, p. 81).

These remedies are possible in tests with different objects (see 10.A on tests of preference). However, they clearly cannot be used when the order effect is due to the continuing effect of experience on the subject under a standard set of conditions, as in the case of the chick avoiding a novel object. The problem of order effects is most acute in studies of development, since studying developmental change often requires measuring the same individual's behaviour repeatedly as it grows older (see following section).

D Studying development

Studies of behavioural development are concerned with describing the changes that occur as an individual matures, and with analysing the processes involved in such changes.

Measuring the behaviour of young animals and children raises some special problems because the organisation of their behaviour alters as they develop. Activities which may look the same at different ages may be controlled in different ways and may have different functions. For example, the amount of time a young rhesus monkey spends in contact with the ventral surface of its mother is influenced primarily by its need for milk early in life, and by its need for a refuge from danger later. Some activities, such as suckling in young mammals, are special adaptations to an early phase and drop out of the repertoire as the individual becomes nutritionally independent of its mother. In general, it is worth recognising that the control of behaviour patterns and their functions are likely to change profoundly as development proceeds.

Behavioural development can be studied by **cross-sectional** research, which involves measuring different individuals at each age, or by **longitudinal** research, which involves measuring the same individuals repeatedly.

One problem with longitudinal studies is that age-related developmental changes and experience of the test situation are inevitably confounded if the same subjects are repeatedly tested as they grow older. For example, testing a young animal's responsiveness to a stimulus can influence its development and thereby affect its behaviour in subsequent tests of responsiveness. Thus, the general problem of order effects due to repeated testing of the same subjects is particularly important in developmental studies (see previous section).

Cross-sectional measurements are not subject to order effects, but do have a number of other drawbacks. First, individuals of two different ages may differ from each other in ways that are not merely due to differences in age and the experience obtained between the two ages at which measurements are made. The separate age groups may have differed prior to the age when the first group was measured. This could arise, for instance, if food availability had a marked influence on the overall pattern of development and fluctuated from one year to the next.

Another problem is that cross-sectional measurement necessarily averages the scores of individuals that may be developing in markedly different ways. This point is illustrated in Fig. 2.1. Here, a measure (*Y*) increases sharply over a narrow age-range, but the timing of this increase varies between individuals. A cross-sectional study, which measured different subjects at each age, would show an apparently gradual increase in *Y*, a pattern of development that would not represent any individual. To take a real example, levels of the hormone testosterone increase markedly in human males over a relatively short period (12–18 months) at the time of puberty, but the age at which this increase starts varies considerably between individuals over a range of several years. A cross-sectional study would give the false impression that testosterone increases gradually over a period of several years. Unfortunately, collecting data longitudinally from long-lived species can take a prohibitive amount of time.

Fig. 2.1. A comparison between longitudinal and cross-sectional measurement. Graph (*a*) shows how a variable (*Y*) changes as a function of age for four individuals, who were measured repeatedly across time (longitudinal measurement). Each individual exhibits an abrupt increase in *Y*, but the age at which this starts differs between the individuals. Graph (*b*) shows the mean score for the same four subjects. This increases gradually as a function of age: a pattern of development that does not represent any individual. A similar picture would have emerged if a different sample of subjects had been measured at each age (cross-sectional measurement). A real example is the sharp increase in the hormone testosterone that occurs at the time of puberty.

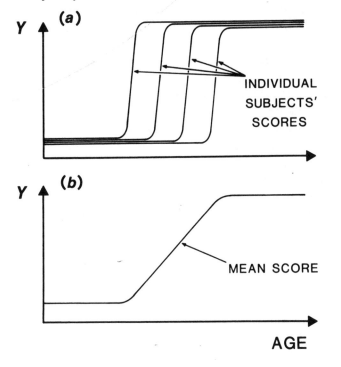

It will be apparent that both cross-sectional and longitudinal approaches have their advantages and both raise different problems of practice and interpretation. Ideally, both methods should be used, as exemplified by the study of the development of pecking in domestic chicks by Cruze (1935). He kept the chicks in the dark from when they hatched and fed them by hand. Starting at different ages, he tested the accuracy with which the chicks pecked at seeds. Once a chick had been tested, it was retested on subsequent days of its life. In this way Cruze was able to obtain cross-sectional data on chicks that were first tested at a given age, and longitudinal data on chicks that were retested each day. Not surprisingly, he found that both a chick's age and its prior experience of pecking at seeds affected its accuracy.

A **sensitive period** in development is an age-range when particular events are especially likely to affect the individual's development (see Bateson, 1979a). The experimental procedures needed to establish that a given condition is more likely to affect subsequent behaviour at one age rather than at others are shown in Fig. 2.2. The age-range when a group of subjects is exposed to the condition is shown by the heavy line. If the exposure were started at different ages and ended at the same age, then the age at first exposure would be confounded with duration of exposure. Thus, any observed effect could have arisen because the individuals that were exposed at an earlier age were also exposed for longer. (Unfortunately, this possibility is not hypothetical; many examples of the ambiguity can be found in the literature.)

A more subtle difficulty is raised by the time of testing. If the time from the end of exposure to testing is not kept constant, then some of the differences between groups could have arisen because the effects of exposure had more time in which to decline in the groups first exposed at the younger ages. However, if time from exposure to testing is kept constant (as shown in Fig. 2.2), the ages and intervening experiences of the groups at testing are necessarily different. Here again, the counsel of perfection is to use both methods for fixing the time of testing.

When experiments are not feasible, the existence of sensitive periods may sometimes be suggested by correlational methods. One approach is to study development **retrospectively**, looking backwards in time to what may have been important events in development. In contrast, a **prospective** approach identifies all the individuals that have had a particular experience and examines what subsequently happens to them. For example, in a restrospective study Bowlby (1951) found that socially maladjusted people were likely to have been deprived of contact with their mother at an early stage in their lives. However, even if a certain type of experience is strongly associated with a particular outcome later in life, a retrospective study cannot detect individuals who received the same early experi-

Fig. 2.2. An experimental procedure for determining the boundaries of a sensitive period in development (a stage in development during which an agent or event, such as a particular type of experience, is particularly likely to have an effect). Exposure to the agent is denoted by the thick horizontal bars and subsequent testing is denoted by the vertical arrows. In this example, five different groups of subjects are exposed to the agent starting at different ages. The duration of exposure is the same for all groups, as is the interval between the end of exposure and testing. However, this procedure means that the groups are tested at different ages.

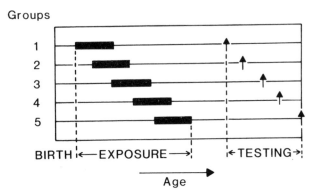

ence but showed no subsequent effect. In fact, such individuals were found to exist in prospective studies (e.g., Douglas, 1975) which showed that maternal deprivation early in life tended to be associated with later behavioural abnormalities in people who came from backgrounds with other social and psychological disadvantages.

E Independence of measurements

Statistical tests generally assume that the data analysed consist of a random sample from the population and that individual data points are statistically independent of one another. This assumption of independence is frequently violated in behavioural research.

1 *The 'pooling fallacy'.* A common error in behavioural research is to treat repeated measurements of the same subject as though they were independent. This practice has been called the 'pooling fallacy' by Machlis *et al.* (1985), who show how it erroneously assumes that the aim of research is to obtain large samples of measurements, rather than measurements from large samples of subjects. The crucial point is that obtaining additional measurements from the same subjects is not a substitute for increasing the number of subjects in the sample.

The pooling fallacy is best illustrated with a hypothetical example given by Machlis *et al.* (1985). Suppose a scientist wishes to estimate the average stride length of adult male cheetahs. Now, measuring the stride length of a cheetah is time-consuming, difficult and costly because it must be done in the laboratory. Furthermore, the sample of individuals available is necessarily small. In order to increase the sample size, therefore, the scientist measures as many strides as possible for each cheetah. Eventually the scientist acquires 100 measurements ($k = 100$) for each of 10 adult male cheetahs ($n = 10$) and pools them to give 1000 measurements ('N' = nk = 10 $\times$ 100). In fact, the scientist has fallen foul of the pooling fallacy. The true sample size (n) is 10, not 1000. The error becomes more obvious if we substitute, say, weight for stride length in this example. Weighing 10 individuals 100 times each and treating the sample size as though it were 1000 rather than 10 would clearly be wrong. Yet the same error is regularly committed in the analysis of behavioural data.

Machlis *et al.* (1985) have used computer simulations to explore the statistical consequences of pooling data. The simulations showed that when data are pooled the probability of incorrectly rejecting a true null hypothesis (that is, finding an apparent effect when none exists; see 9.C.3) is generally higher than the assumed level of significance and, in many cases, may exceed 0.50. In other words, an effect that appears to be statistically significant at, say, the 0.05 level may in fact have arisen through sampling error with a probability of more than ten times this level.

The pervasiveness of pooling in behavioural research is convincingly demonstrated by Machlis *et al.* (1985). They surveyed the articles published in two major behavioural journals and found that at least 20% of articles in one journal and 52% in the other contained pooled data. In addition, they point to examples of pooled data sets in books devoted to behavioural research methods and in their own published work (e.g., Machlis, 1977).

In general, then, repeated measurements from the same subject should be averaged to give a single data point for that subject and the sample size (n) should be equal to the number of *subjects*, not the number of measurements.

It is not always necessary to throw away information about variation for a single individual, particularly when all individuals have been measured approximately the same number of times. Statistical methods such as analysis of variance (ANOVA) can be used to analyse within-subject variation as well as sources of variation between subjects.

2 'Litter effects' and other group effects. Problems with independence also arise when studying species that have multiple offspring – for example, rats or cats – because of likely correlations between measurements on individuals from the same litter, or 'litter effects'. If within-litter correlations are ignored, the statistical significance of differences between groups of subjects is likely to be over-estimated

(Abbey & Howard, 1973). For example, suppose six rat pups from the same litter are weighed, having been designated as the experimental group, and six pups from another litter are designated as the control group. If by chance the mother of the experimental group happened to be large, then all her pups might also tend to be large, thereby confounding a litter effect with the treatment effect. The sample size of each group should be one (litter) and not six (pups).

Thus, if within-litter variation is significantly smaller than between-litter variation, measurements of littermates cannot be considered to be statistically independent. One solution is to measure only one randomly selected individual from each litter, although this procedure throws away potential information. A better alternative is to measure all the littermates, but to treat the *litter-mean* value as a single measurement, so that the sample size (n) is equal to the number of litters rather than the number of individuals (Abbey & Howard, 1973). A third approach is to use analysis of variance (ANOVA) to pick out 'litter effects'. This requires a balanced experimental design, in which one or more members of each litter (preferably equal numbers) are included in each treatment group, so that the effects of the experimental treatment and between-litter variation are not confounded.

Problems of non-independence are not restricted to similarities between littermates, and apply to any grouping of subjects that can give rise to systematic within-group correlations. For example, animals housed in the same cage may be more alike than animals from other cages (so-called cage effects). In studies of social behaviour, interactions between members of the same group can lead to problems of independence, either through short-term effects such as social facilitation or, in the long term, through factors such as using a common home range. In many field studies of social species there are good grounds for arguing that *groups* rather than individuals should be treated as the independent units of measurement.

In all cases where there are doubts about the independence of individuals in a group, such as littermates, occupants of the same

cage or members of the same social group, the safest policy is to use
the mean value for the group as a single measurement.

3 *Independence of categories.* Another common problem arises when
different measures or categories are not independent of one another.
This can cause problems in interpreting associations between measures. For example, two measures can be correlated either because
there is a non-trivial association between the two types of behaviour
or because the two measures are merely two ways of measuring the
same thing. For instance, suppose two mutually exclusive categories
of behaviour are found to be negatively correlated. Now, if one
category (A) accounts for a considerable proportion of the observation time, then the negative correlation may have arisen simply
because there was no time for the other type of behaviour (B) to
occur. In effect, the second category (B) is equivalent to 'not-A'.
Two mutually exclusive categories of behaviour that together account
for a large proportion of total observation time are bound to be
negatively correlated and any such correlation must be interpreted
with caution.

When using different types of measure to assess the same behaviour
pattern, it is important to ensure that the various measures are,
in principle, independent of one another. For example, suppose a
particular type of behaviour is measured in three ways: in terms of
its mean duration; total number of occurrences; and total duration.
Now, these three measures are not independent, since mean duration
is equal to the total duration divided by the number of occurrences.
Given any two of these measures the third can be calculated, therefore
only two measures can be regarded as independent descriptions of
the behaviour (see Cane, 1961; Sackett *et al.*, 1978).

F How much information to collect

Up to a point, the more data collected the better, because
statistical power (see 9.C.4) is improved by increasing the sample
size. However, the temptation to carry on collecting results indefin-
itely must be weighed up against the time involved, since at some
stage it will be more productive to move on to a new study than to

keep collecting additional data for the current one. Increasing the sample size by amassing more and more results eventually offers diminishing returns in terms of increased statistical power. In other words, when sufficient results have been acquired, additional results may add little to the ability to draw conclusions.

1 *Internal consistency.* A crude but simple way of checking whether sufficient data have been collected is to divide the data randomly into two halves and analyse each half separately. If the two sets of data both generate clear conclusions that are in agreement, then sufficient results have probably been obtained. If, however, the two sets of data lead to conflicting conclusions, or are insufficient to produce any firm conclusions, then more data are almost certainly needed.

A more sophisticated approach, known as **split-half analysis**, is to divide the data for a particular category of behaviour randomly into two halves and calculate the correlation between the two sets of data. If the correlation coefficient is sufficiently high (say, $r > 0.7$), then the data set is said to be reliable (see Ghiselli *et al.*, 1981, p. 232; and 6.C on measuring reliability using correlation).

2 *Estimating the necessary sample size.* The aim of some studies is to estimate a population parameter (see 9.C.6); for instance, the mean vocalisation rate of a species under certain conditions, or the mean concentration of a hormone in the bloodstream. The sample size needed to give a sample mean that is within specified limits of the true (population) mean value can be estimated, provided some pilot measurements are available to give an estimate of the population standard deviation.

To estimate the necessary sample size (n), three pieces of information are required: first, an estimate of the population standard deviation (σ); second, the level of statistical significance to be attached to the estimate (α); and third, the maximum acceptable difference (D) between the sample mean and the true (population) mean (i.e., $D = |\bar{X} - \mu|$). The minimum sample size required (n) is given by the following relation (Harnett, 1982, p. 329):

$$n = \frac{\sigma^2 z_{\alpha/2}^2}{D^2}$$

(where $z_{\alpha/2}$ is the critical value of the cumulative normal variable [z] at the $\alpha/2$ level of significance).

For example, suppose an observer wishes to estimate the mean frequency with which infant rhesus monkeys of a certain age approach their mothers. A pilot study, comprising six 1-h periods of observation, recorded the following independent measurements of approach rate (measured in number of approaches per hour): 9, 12, 7, 15, 13, 4 h^{-1}. From this pilot sample, the population standard deviation (σ) is estimated to be 4.10 h^{-1}. Suppose the investigator wishes the maximum difference (D) between the eventual sample mean and the true (population) mean to be no more than 2 h^{-1} and the level of significance (α) is set at 0.05. (In other words, the observer wishes to be 95% confident that the sample mean will be within 2 h^{-1} of the true value.) A table of critical values for the normal variable z is used to find z at the 0.025 ($= \alpha/2$) level. This shows that $z\,(0.975) = 1.96$. Hence, the required sample size (n) is calculated as:

$$n = (4.10^2 \times 1.96^2)/2^2 = 16.1$$

Thus, at least 17 independent measurements of approach rate are needed in order to be 95% confident that the measured (sample) mean rate is within 2 h^{-1} of the true (population) mean value. (Note that n is always rounded up.)

Of course, the information needed to estimate the required sample size is not always readily available and, even if it is, the required sample size may be larger than is practically possible. The larger the sample size the greater the statistical power (see 9.C.4) and acceptability of a study, but the less feasible it becomes in practice.

The necessary sample size can also be estimated in situations other than determining a parameter (see Cohen, 1977).

G **When to observe**

Making observations according to some predetermined schedule is an important precaution against the bias that would arise if the observer merely recorded whenever something obvious or interesting happened. In general, the time at which the recording session starts and stops should be determined in advance, and not by what the subjects are doing at the time (unless, of course, the aim of the study is to find out what happens during or after a particular type of behaviour has occurred; see 4.A.4).

Choosing the appropriate time of day at which to observe is an important practical issue in any study. Obviously, animals are not equally active throughout the 24-h period, so the amount of activity seen will depend on the time of day at which the subjects are watched. A more subtle point is that not just the overall amount, but also the *nature* of behaviour may change according to the time of day. For example, a field study of gorillas found that the nature of their social interactions varied according to their overall level of activity and, therefore, according to the time of day at which they were watched (Harcourt, 1978).

This problem can be approached in one of four ways:

(*a*) By recording behaviour throughout the 24-h period, either by continuous observation or with several observation sessions spread across each day. Clearly, this is not a practical proposition in many cases, particularly if only one observer is involved. A compromise might be to observe two or three times each day; for example, during the morning and early evening. If the results obtained at the various times of day are markedly different then they must be analysed and treated separately; if not, they can be pooled to give a daily average.

(*b*) By observing at a different time on each day such that, averaged across the entire study, each part of the day is equally represented in the final sample (e.g., Harcourt & Stewart, 1984). This approach cannot be used if there is any likelihood that the behaviour changes systematically from day to day – as, for example, when studying young, developing animals – or when behaviour undergoes marked seasonal changes. Despite good intentions of sampling uniformly throughout the day, many investigators find that their samples are in practice unevenly distributed.

(*c*) By partially ignoring the problem and observing at the same time each day. This is the most usual approach, especially in laboratory studies. Strictly speaking, if all observations are made at the same time of day then the results should not be generalised to any other time of day. This limitation on the validity of observations should not present great problems unless the time of day strongly influences the nature of the results – particularly if the aim of the study is to make comparisons between groups of subjects from the same species. Problems can, however, arise when activity rhythms drift or when comparing the behaviour of different species whose activity rhythms differ. Observations should be made at a time of day when the behaviour of interest is most likely to be occurring. Obviously, studying social behaviour at a time of day when the animals are usually asleep is pointless.

(*d*) By ignoring the problem completely and recording at a different time each day on a haphazard basis. This approach has no obvious merits, but is sometimes a necessary evil for observers studying behaviour under difficult conditions (see Chapter 7 on field studies).

H **Floor and ceiling effects**

Two groups of subjects may appear not to differ, when in reality they do, because all the scores are clumped at one or other end of the possible range of values. Genuine differences will be obscured if all or most subjects obtain the minimum possible score (a **floor effect**) or the maximum possible score (a **ceiling effect**). For example, a test of mathematical ability involving multiplication by two is unlikely to reveal differences between human adults because most people will answer all the questions correctly (a ceiling effect). Clearly, a more difficult test is more likely to reveal differences, but a test that was too hard would result in most people answering none of the questions correctly (a floor effect). Although this point may seem obvious, it is often overlooked as a possible explanation when negative results are obtained; for example, when two groups of subjects are found not to differ significantly. Floor and ceiling effects apply to correlations as well as differences, since two measures will appear to be uncorrelated if either set of scores is clumped at one end of its range of measurement.

Examination of preliminary data should reveal whether floor or ceiling effects are likely (see 3.A and 9.B). If they are, matters may be improved by choosing a different, related measure which produces a broader spread of scores. For example, a successive choice test (see 10.A) that simply measures whether or not two objects are approached may seem to show that both objects are equally attractive. However, a more sensitive measure of preference, such as latency to approach, may reveal clear differences in the effectiveness of the two objects (e.g., Bateson, 1979*b*).

I **Further reading**

Experimental design in biology and the behavioural sciences is considered by Lehner (1979), Clarke (1980) and Sprinthall (1982, ch. 10). See also Medawar (1981) for a general discussion of scientific methods. Experimenter effects are reviewed by Rosenthal (1976) and Rosenthal & Rubin (1978); see also Hollenbeck (1978). The methodological problems of developmental work are examined by Wohlwill (1973). Bateson (1979*a*) discusses sensitive periods. Independence and the 'pooling fallacy' are reviewed by Machlis *et al.* (1985); see also Cane (1961), Abbey & Howard (1973) and Sackett *et al.* (1978). The problem of how activity patterns and time of day affect what is seen during observations is considered by Harcourt (1978) and Simpson (1979).

3

Preliminaries to measuring behaviour

A Preliminary observation

Quantitative recording of behaviour should be preceded by a period of informal observation, aimed at understanding and describing both the subjects and the behaviour it is intended to measure. Preliminary observation is important for two reasons: first, because it provides the raw material for formulating questions and hypotheses; and second, because choosing the right measures and recording methods requires some knowledge of the subjects and their behaviour. Preliminary observation is especially important if the problems or animals are new to the investigator.

The hypotheses that direct a study can seldom be formulated in isolation; rather, they reflect existing knowledge and theories, as well as the investigator's own interests and hunches. Effective research therefore requires the investigator to be familiar with the subjects, both by direct experience of watching them and by reading the literature about their biology and behaviour. A period of preliminary observation also provides a valuable opportunity for sharpening up questions and hypotheses, practising recording methods and generating additional or supplementary hypotheses. Most biologists would argue that an initial phase of description is an essential precursor to quantification, since before the right questions can be asked it is essential to know what there is to ask questions about.

We cannot over-state the importance of simply watching before starting to measure systematically. Beginners often get bogged down early in a study because they rush to obtain 'hard data' and do not allow sufficient time to watch, think and frame interesting questions. Even an experienced observer needs to spend time on preliminary observation.

As a practical guideline, we suggest that investigators should always *plan* not to include data obtained during the first few recording sessions in their final analysis. Otherwise it can be tempting to use all the data that have been obtained, even though data from early recording sessions may be unreliable (see 6.A), or not comparable to later data because of **observer drift** (see 6.E) or deliberate changes in measurement procedures. After a period of trial recording sessions, in which behavioural categories and measurement techniques have been tried out, preliminary data should be analysed. It is at this stage that hypotheses and methods should be modified if necessary.

B Describing behaviour

Behaviour can be described in a number of different ways. The simplest distinction is between describing behaviour in terms of structure or consequences.

1 *The structure* is the appearance, physical form or temporal patterning of the behaviour. The behaviour is described in terms of the subject's posture and movements.

2 *The consequences* are the effects of the subject's behaviour on the environment, on other individuals or on itself. In this case, behaviour may be described without reference to how the effects are achieved. Categories such as 'obtain food' or 'escape from predator' are described in terms of their consequences, and may be scored irrespective of the actual pattern of body movements used.

For example, 'turn light on' is a description in terms of consequences, while 'press switch down using index finger' is a structural description.

Similarly, 'run tip of bill along primary feather of wing' is a structural description, while 'preen' is a description by consequence.

Describing behaviour by its structure can sometimes generate unnecessary detail and place demands on the observer's ability to make subtle discriminations between complex patterns of movement. Description by consequence is often a more powerful and economical approach, and has the additional advantage that the consequences can sometimes be recorded using automatic devices (see 5.A).

A third form of description is in terms of the individual's spatial **relation** to features of the environment or to other individuals. In this case, the subject's position or orientation relative to something (or someone) is the salient feature. In other words, the emphasis is not on what the subject is doing, but where or with whom. For example, 'approach' or 'leave' might be defined in terms of changes in the spatial relation between two individuals.

It is not uncommon for behaviour to be described in terms of presumed consequences or causes which later turn out to be wrong. Because of this danger it is best to use neutral terms for labelling behaviour patterns, rather than labels which falsely imply knowledge of the animal's internal state or the biological function of the behaviour pattern. For example, if a category of vocalisation is named *distress call* (rather than given a neutral label, such as *peep*), then an observer might be tempted to include vocalisations which did not meet the stated criteria for the category, but which were emitted when the animal was apparently distressed.

C   Choosing categories

Behaviour consists of a continuous stream of movements and events. Before it can be measured, this stream must be divided up into discrete units or categories. In some cases behaviour appears to be composed of 'natural units' of clearly distinguishable, relatively stereotyped behaviour patterns (such as a peck), and the division process will partly be dictated by the behaviour itself. (Highly stereotyped, species-characteristic behaviour patterns were referred to by early ethologists as 'fixed action patterns'. However, the term 'modal action pattern', or just 'action pattern', is now preferred since even these types of behaviour pattern do show some variability: see Barlow, 1977; Bekoff, 1977; Dawkins, 1983.) To a large extent, though, the suitability of a category depends on the question that is being asked rather than on some inherent feature of the behaviour. Indeed, as we emphasised in Chapter 1, observational categories must reflect some sort of implicit theory and do not have an existence of their own, independent of the observer. It is difficult, therefore, to give specific advice on what sorts of categories to choose, although we can offer some general guidelines.

(*a*) Enough categories should be used to describe the behaviour in sufficient detail to answer the questions and, preferably, to provide some additional background information.

(*b*) Each category should be precisely defined (see 3.D) and should summarise as much relevant information as possible about the behaviour.

(*c*) Categories should generally be independent of one another; that is, two or more categories should not formally be different ways of measuring the same thing (see 2.E and Cane, 1961).

(*d*) Categories should generally be homogeneous; that is, all acts included within the same category should share the same properties.

Inexperienced observers often err on the side of trying to record too much. A given stream of behaviour could potentially be described in an almost limitless number of ways, depending on the questions being asked, so it is essential to be somewhat selective. It is certainly best to drop categories that are clearly irrelevant, or which seem inconsistent and difficult to measure reliably (see 6.A). The chances are that the fewer categories used, the more reliably each will be measured. Bear in mind, though, that observers do improve with experience, so data from later recording sessions may be reliable even if data from early sessions are not. Furthermore, there are circumstances in which it is better to record too much initially, rather than regretting it later. Redundant or unreliable categories can always be discarded or pooled at the analysis stage. It is also wise to collect supplementary information which might in the future provide useful background or raise new questions. However, collecting a wide range of measures or supplementary information should not be allowed to detract from careful measurement of the important things.

The extent to which the definitions of individual categories are specific rather than general will depend on the nature of the problem. Questions and hypotheses tend initially to be rather broad and then narrow down as more is discovered about a particular problem. The more clearly and precisely the initial question has been formulated, the more obvious it will be what to measure.

When choosing categories it can sometimes be helpful to have descriptions of the main types of behaviour pattern that typify the species. In some cases this information is available in the form of an **ethogram**, which is a catalogue of descriptions of the discrete, species-typical behaviour patterns that form the basic behavioural repertoire of the species. Unfortunately, published ethograms vary enormously in the number of behavioural categories included and the detail with which these are described, and ethograms are unavailable for many commonly studied laboratory subjects (see Schleidt *et al.*, 1984). Moreover, ethograms are of limited use because not all members of a species behave in the same 'species-typical' way. On the contrary, individuals of the same species can behave in quite different ways (see Slater, 1981; Davies, 1982). Look up.

D **Defining categories**

Each category of behaviour to be measured should be clearly, comprehensively and unambiguously defined, using criteria that can be easily understood by other observers. More important still, the criteria used to define a category should unambiguously distinguish it from other categories, particularly those it resembles most closely. A detailed and complete definition of each category and the associated recording method should be written down *before* the data used in the final analysis are collected.

The period of preliminary observation provides an opportunity to develop the precise criteria used to define each category. A completely satisfactory and unambiguous definition of a category can rarely be formulated without having watched the behaviour for some time. Preliminary definitions are often unable to deal with unforeseen

ambiguous examples of the behaviour that crop up during preliminary observations, and must be therefore be modified in the light of
experience.

Clearly, all the data for a particular category that are used in the
final analysis must be comparable. Thus, data obtained before the
final definition of a category was formulated must be discarded.
Developing a set of precise and unambiguous category definitions
can be a slow process.

Writing down precise definitions of categories at the beginning of
the study is essential to prevent definitions and criteria from 'drifting'
during the course of the study (see 6.E on factors affecting reliability).
Written definitions should be sufficiently precise and detailed to
enable another observer to record exactly the same things.

E Types of measure: latency, frequency, duration and intensity

Behavioural observations most commonly yield four basic
types of measure.

1 *Latency* (measured in units of time; e.g., s, min or h) is *the
time from some specified event* (for example, the beginning of the
recording session or the presentation of a stimulus) *to the onset of
the first occurrence of the behaviour*. For example, if a rat presses
a lever for the first time 6 min after it is placed into a Skinner box,
the latency to press the lever is 6 min.

2 *Frequency* (measured in reciprocal units of time; e.g., s^{-1}, min^{-1} or
h^{-1}) is *the number of occurrences of the behaviour pattern per unit
time*. Frequency is a measure of the *rate* of occurrence. For example,
if a rat presses a lever 60 times during a 30-min recording session,
the frequency of lever pressing is 2 min^{-1}.

An alternative usage, which is perhaps more common in the
behavioural literature (and in statistics), is when 'frequency' refers
to the *total number of occurrences*. However, this is uninformative
and potentially misleading unless the total time for which the
behaviour was watched is also specified. For example, to state that

the 'frequency' of a behaviour was 60 is meaningless: did it happen 60 times in two minutes; one hour; 3.7 hours; a day...? Most statements about total numbers of occurrences could equally well refer to rates of occurrences, since a total number of occurrences can always be expressed as a rate (assuming the length of the observation period is known). To avoid any confusion, the total number of occurrences should be explicitly referred to as such. Expressing frequencies in the way we suggest (number per unit time) removes any possible ambiguity.

A further ambiguity in the use of the term frequency arises in studies of the acoustic properties of vocalisations, when frequency is used in the sense of pitch (measured in Hz, or cycles per second: see Appendix 3, Table a3.1). In such studies, ambiguity is best avoided by using 'frequency' with its acoustic meaning and 'rate of occurrence' to refer to the incidence of behaviour patterns.

3 *Duration* (measured in units of time; e.g., s, min or h) is *the length of time for which a single occurrence of the behaviour pattern lasts*. For example, if a kitten starts suckling and stops 5 min later, the duration of that period of suckling was 5 min.

'Duration' is also used in at least two other senses in the behavioural literature. The first is when 'duration' (or 'total duration') refers to the *total* length of time for which all occurrences of the behaviour lasted over some specified period, usually the whole observation session. A total duration is, of course, meaningless unless the total time for which the behaviour was watched is also specified. For example, to state that the 'total duration' of a behaviour pattern was 16 min says nothing: was it 16 min out of 20 min; 30 min; an hour; a day...? To avoid any ambiguity, we recommend that a total duration should be expressed as the total duration over the specified period of observation (for example, '9 min per 30 min') and should be explicitly referred to as **total duration**.

Alternatively, a total duration can be expressed as a proportion (or percentage) of the observation period, in which case it should be explicitly referred to as the **proportion** (or percentage) **of time spent performing the behaviour**. For example, if a kitten spent a total of

10 min suckling during a 30-min observation session, then the proportion of time spent suckling was $10/30 = 0.33$. Note that a proportion of time is a dimensionless index with no units of measurement.

Expressing a duration as a proportion or percentage of total time omits the potentially important information about the total time for which the behaviour was watched. For example, the interpretation placed on the statement that the proportion of time spent sleeping by a subject was 0.10 must depend on whether this figure was based on, say, a 24-h period of observation as opposed to a 30-min observation.

'Duration' (or 'mean duration') is also used to refer to the *mean* length of a single occurrence of the behaviour pattern, measured in units of time (e.g., s, min or h). This is obtained by recording the duration of each occurrence of the behaviour pattern and calculating the mean of these durations. To avoid any possible ambiguity, we suggest that this measure should be referred to explicitly as a **mean duration**. Mean duration can also be calculated by dividing the total duration of the behaviour pattern by the total number of occurrences. (This has the advantage that the duration of each occurrence need not be recorded separately; the observer could, for example, use a cumulative stopwatch to record the total duration and a counter to

Fig. 3.1. The meaning of latency, frequency and duration. The black bars represent three successive occurrences of a behaviour pattern during an observation period of length t units of time. The *durations* of the three occurrences are, respectively, a, b and c units of time. The *total duration* is $(a + b + c)$ units of time. The *mean duration* is $(a + b + c)/3$ units of time. The *proportion of time spent* performing the behaviour pattern is $(a + b + c)/t$. The *frequency* is $3/t$ per unit time. The *total number of occurrences* is 3.

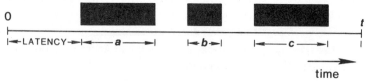

record the total number of occurrences.) According to one definition of the term 'bout', the mean duration of a behaviour pattern is equivalent to its **mean bout length**. However, bout length can be defined in other ways (see10.B).

As an illustration of these various measures, suppose that a mother and infant are observed for 60 min, during which suckling occurred five times; the individual periods of suckling lasting 3 min, 10 min, 1 min, 1 min and 1 min, respectively. According to our suggested definitions, the *durations* of suckling were 3,10,1,1 and 1 min; the *total duration* of suckling was 16 min per 60 min; the *proportion of time spent suckling* was 0.27 (= 16/60); and the *mean duration of suckling* was 3.2 min (= 16/5).

Frequency and duration, which are the measures most commonly used for describing behaviour, can give different and complementary pictures. For example, how often two monkeys groom each other (frequency) tells us something different about the nature of their social relationship from how long they spend doing it (duration), since frequency describes the initiation of grooming while duration describes its continuation. In fact, empirical studies have shown that frequency and duration measures of the same behaviour are not always highly correlated, implying that it is probably wise to record both (Dunbar, 1976; Rhine & Flanigon, 1978; Rhine & Linville, 1980).

4 *Intensity*. In general, categories are best defined in such a way that the behaviour is simply scored according to whether or not it has occurred or for how long it has occurred, rather than making assessments of intensity or amplitude. Intensity, unlike latency, frequency and duration, has no universal definition. Nonetheless, it may be helpful or even essential to make judgements about the intensity or amplitude of a behaviour pattern. For example, it may be important to measure the sound intensity of a vocalisation, the amplitude of a limb movement or the height of a jump; or to estimate the intensity of a facial expression or the aggressiveness of a social interaction (see 10.D on ratings). Intensity can sometimes be measured according to the presence or absence of certain components of the act, which may be present at high intensity but absent at low intensity (Slater, 1978).

A simple and informative index of intensity is **local rate**, a measure originally used in operant psychology. The local rate of an activity (such as eating, walking or grooming) is defined as *the number of component acts per unit time spent performing the activity* (Roper, 1984). For example, suppose that an activity – eating – is composed of discrete, component acts – the ingestion of individual food pellets. The local rate of eating would, in this case, be given by the number of pellets ingested per unit time spent eating (see Fig. 3.2). Similarly, the intensity of walking might be measured by the number of strides per unit time spent walking, and the intensity of grooming by the number of face-stroking paw movements per unit time spent grooming. Local rate captures the speeded-up or hurried nature of intense behaviour: the more hurriedly an activity is performed, the higher its local rate.

Finally, in some cases the consequences of behaviour can be measured in terms of some physical quantity related to the behaviour, such as the weight of food eaten, the volume of water drunk, the number of prey captured or the distance travelled.

Fig. 3.2. Local rate as a measure of intensity. An activity (for example, eating) is composed of discrete, component acts (the ingestion of individual food pellets), indicated by the vertical lines. The local rate is given by the total number of occurrences of the component act (number of pellets eaten during the observation session) divided by the total duration of the activity (total time spent eating = $t_1 + t_2 + t_3 + ... + t_n$). (After Roper, 1984.)

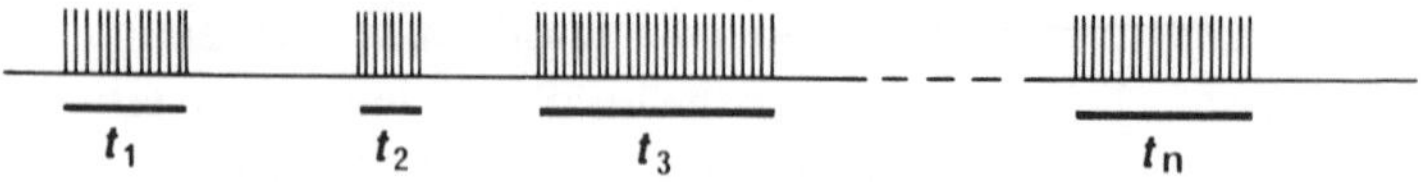

$$\text{Local rate} = \frac{\text{Total no. of occurrences}}{(t_1 + t_2 + t_3 + ... t_n)}$$

F   Events versus states

When choosing the type of measure used to describe a behaviour pattern, it is helpful to distinguish between two fundamental types of behaviour pattern, which lie at opposite ends of a continuum.

1 *Events* are behaviour patterns of relatively short duration, such as discrete body movements or vocalisations, which can be approximated as points in time. The salient feature of events is their *frequency* of occurrence. For example, the number of times a dog barks in 1 min would be a measure of the frequency of a behavioural event.

2 *States* are behaviour patterns of relatively long duration, such as prolonged activities, body postures or proximity measures. The salient feature of states is their *duration* (mean or total duration, or the proportion of time spent performing the activity). For example, the total time a dog spends asleep over a 24-h period would be a measure of the total duration of a state. (Note that the term 'state' is also used in the behavioural literature to refer to a motivational state, such as hunger or thirst, so it is important not to confuse the two.)

The onset (or termination) of a behavioural state can itself be scored as an event and measured in terms of its frequency.

G   Further reading

Many of the basic issues concerned with describing, classifying and measuring behaviour are dealt with by Hinde (1970, ch. 2; 1982, ch. 1), Hutt & Hutt (1970, ch. 3), Slater (1978) and Lehner (1979). The 'units of behaviour' problem is considered by Barlow (1977), Purton (1978), Slater (1978) and Dawkins (1983). Altmann (1974) discusses sampling rules and the distinction between events and states in her paper on observational methods.

4

Recording methods

A Sampling rules: *ad libitum*, focal, scan and behaviour sampling

When deciding on systematic rules for recording behaviour, two levels of decision must be made. The first, which we refer to as **sampling rules**, specifies which subjects to watch and when. This covers the distinction between *ad libitum* sampling, focal sampling, scan sampling and behaviour sampling. The second, which we refer to as **recording rules**, specifies *how* the behaviour is recorded. This covers the distinction between continuous recording and time sampling (which, in turn, is divided into instantaneous sampling and one-zero sampling; see Fig. 4.1). In this section, we consider the four different sampling rules.

1 *Ad libitum sampling* means that no systematic constraints are placed on what is recorded or when. The observer simply notes down whatever is visible at the time and seems relevant. Clearly, the problem with this method is that observations will be biased towards those behaviour patterns and individuals which happen to be most conspicuous. Provided this important limitation is borne in mind, however, *ad libitum* sampling can be useful during preliminary observations, or for recording rare but important events.

2 *Focal sampling* (or 'focal animal sampling') means observing one individual (or one dyad, one litter or some other unit) for a specified amount of time and recording all instances of its behaviour – usually for several different categories of behaviour. Ideally, the choice of focal individual is determined prior to the observation session. When recording the social behaviour of the focal individual it may also be necessary to record certain aspects of other individuals' behaviour, such as who initiates interactions and to whom behaviour is directed. Focal sampling is generally the most satisfactory approach to studying groups.

Inevitably the focal individual will, on occasions, be partially obscured or move completely out of sight, in which case recording has to stop until it is visible again. Any such interruption should be recorded as 'time out', and the final measure calculated according to the time for which the focal individual was visible. Note, however, that omitting 'time out' may introduce bias if subjects systematically tend to do certain things whilst out of sight. For example, many animals seek privacy when eating or mating, so their behaviour when visible to the observer is not necessarily representative of their behaviour as a whole (e.g., Harcourt & Stewart, 1984).

Focal sampling can be particularly difficult under field conditions, because the focal individual may leave the area and disappear completely. If this happens, should it be pursued and if so, for how long; or should the observations be stopped and a new focal individual chosen? Explicit rules are needed for deciding what to do if the focal individual does disappear in the middle of a recording section (see also 7.B).

Note that some authors confusingly use 'focal (animal) sampling' as a synonym for continuous recording (see 4.C below), conflating a sampling rule (who is watched) with a recording rule (how their behaviour is recorded). In fact, any of the three different recording rules (continuous recording, instantaneous sampling or one-zero sampling) can be used when recording the behaviour of a single subject (focal sampling).

3 *Scan sampling* means that a whole group of subjects is rapidly scanned, or 'censused', at regular intervals, and the behaviour of each individual at that instant recorded. The behaviour of each individual scanned is, necessarily, recorded by *instantaneous sampling* (see 4.D, below), although the two terms are not synonymous (see Fig. 4.1). Scan sampling usually restricts the observer to recording only one or a few simple categories of behaviour, such as whether or not a particular activity is occurring, or which individuals are asleep.

The time for which each individual is watched in a scan sample should, in theory, be negligible; in practice, it may at best be short and roughly constant. A single scan may take anything from a few seconds to several minutes, depending on the size of the group and the amount of information recorded for each individual (see Altmann, 1974).

An obvious danger with scan sampling is that results will be biased because some individuals or some behaviour patterns are more conspicuous than others. For example, Harcourt & Stewart (1984) found that previously published figures for the amount of time spent feeding by gorillas in the wild were too low. They argued that this was because previous studies had used scan sampling, and gorillas are less visible to observers when they feed. Focal animal sampling with the subjects continuously in view, which was used by Harcourt & Stewart to obtain a better estimate of time spent feeding, is not subject to this source of bias.

Scan sampling can be used in addition to focal sampling during the same observation session. For example, the behaviour of a focal individual may be recorded in detail, but at fixed intervals (say, every 10 or 20 min) the whole group is scan-sampled for a single category, such as the predominant activity or proximity to each other.

If scan samples are to be used as separate data points, rather than averaged to provide a single score, they must be statistically independent of one another (see 2.E). This means that they must be adequately spaced out over time: clearly, scan samples taken at, say, 30-s intervals would not constitute independent measurements.

Note that some authors use 'scan sampling' to refer to instantaneous sampling (see 4.D below), again conflating a sampling rule with a recording rule.

4 *Behaviour sampling* means that the observer watches the whole group of subjects and records each occurrence of a particular type of behaviour, together with details of which individuals were involved. Behaviour sampling is mainly used for recording rare but significant types of behaviour, such as fights or copulations, where it is important to record each occurrence. Rare behaviour patterns would tend to be missed by focal or scan sampling. Behaviour sampling is often used in conjunction with focal or scan sampling and is subject to the same source of bias as scan sampling, since conspicuous occurrences are more likely to be seen. (Indeed, behaviour sampling is sometimes referred to as 'conspicuous behaviour recording'.)

B   Recording rules: continuous recording versus time sampling
Recording rules are of two basic types:

1 *Continuous recording* (or 'all-occurrences' recording). This method aims to provide an exact and faithful record of the behaviour, measuring true frequencies and durations and the times at which behaviour patterns stopped and started (see 4.C below).

2 *Time sampling.* Here, the behaviour is sampled periodically, therefore less information is preserved and an exact record of the behaviour is not necessarily obtained. Time sampling can be divided into two principal types: **instantaneous sampling** and **one-zero sampling** (see 4.D and 4.E below). The hierarchy of sampling rules and recording rules is represented in Fig. 4.1.

Time sampling is a way of condensing information, thereby making it possible to record several different categories of behaviour simultaneously. In order to do this, the observation session is divided up into successive, short periods of time called **sample intervals**. (The choice of sample interval is discussed in 4.F.) The instant of time at the end of each sample interval is referred to as a **sample point** (see Fig. 4.2). For example, a 30-min observation session might be divided up into 15-s sample intervals giving 120 sample points. Successive sample points are denoted by a stopwatch or, more conveniently, by a **beeper** – a small electronic timer which gives the observer an audio cue through an earphone once every sample interval (see Appendix 2). The distinction between continuous recording and time sampling also applies when computers are used to record data automatically (see Chapter 5).

C **Continuous recording**

With continuous recording (or 'all-occurrences' recording), *each occurrence* of the behaviour pattern is recorded, together with information about its time of occurrence. True continuous recording aims to produce an exact record of the behaviour, with the times at which each instance of the behaviour pattern occurred (for events), or began and ended (for states; see 3.F).

1 *Measures obtained.* For both events and states, continuous recording generally gives true frequencies, and true latencies and

Fig. 4.1. The hierarchy of *sampling rules* (determining who is watched and when) and *recording rules* (determining how their behaviour is recorded).

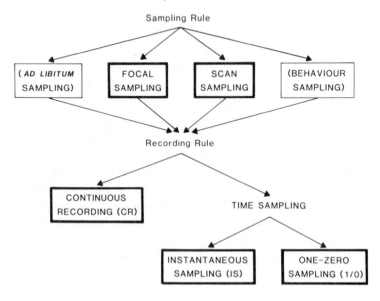

durations if an exact time base is used. However, bias can arise if a measurement of duration or latency is terminated before the bout of behaviour actually ends, either because the recording session ends or because the subject disappears from view. This is because the longer a bout of behaviour lasts, the more likely it is that its duration will be under-estimated by the termination of recording.

2 *Applications.* Continuous recording preserves more information about a given category of behaviour than time sampling, and should be used whenever it is necessary to measure true frequencies or durations accurately. Continuous recording is also necessary when the aim is to analyse sequences of behaviour (see 4.H). However, its use can be limited by practical considerations, because continuous recording is more demanding for the observer than time sampling. One consequence of this is that fewer categories can be recorded at any one time. Trying to record everything can mean that nothing is measured reliably.

In practice, continuous recording is typically used for recording the frequencies of discrete events and for recording the durations of

Fig. 4.2. The division of an observation session into successive, short units of time, or *sample intervals*, for the purposes of time sampling. The end of each sample interval (*sample point*) is often denoted by an audio *beeper*.

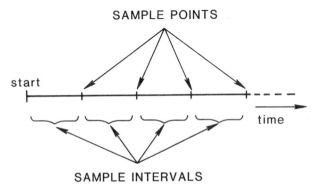

behavioural states, particularly when it is important to preserve information about the sequence of behaviour patterns.

D Instantaneous sampling

With instantaneous sampling (or 'point sampling', or 'fixed-interval time point sampling'), the observation session is divided into short sample intervals. *On the instant of each sample point (i.e., on the 'beep'), the observer records whether or not the behaviour pattern is occurring* (see Fig. 4.3).

1 *Measure obtained.* The score is expressed as the *proportion of all sample points on which the behaviour pattern was occurring.* For example, if a 30-min recording session was divided into 15-s sample intervals, and a behaviour pattern occurred on 40 out of the 120 sample points, then the score would be 40/120 = 0.33. (Note that an instantaneous sampling score is a dimensionless index with no units of measurement.) Instantaneous sampling gives a single score for the whole recording session: obviously, individual sample points within a session cannot be treated as statistically independent measurements.

Instantaneous sampling does not give true frequencies or durations. However, if the sample interval is short relative to the average duration of the behaviour pattern, then instantaneous sampling can produce a record that approximates to continuous recording. The shorter the sample interval, the more accurate instantaneous sampling is at estimating duration, and the more closely it resembles

Fig. 4.3. Instantaneous sampling (or 'point sampling'). The behaviour pattern, denoted here by the black bars, is scored according to whether or not it is occurring on the instant of each successive sample point ('on the beep').

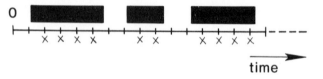

continuous recording. If the sample interval is short, then an instantaneous sampling score gives a direct estimate of the proportion of time for which the behaviour occurred. For example, if the instantaneous sampling score was 0.25, the best estimate of the proportion of time spent performing the behaviour would also be 0.25.

The accuracy of instantaneous sampling depends on the length of the sample interval (which should be as short as possible), the average duration of the behaviour pattern (which should be long relative to the sample interval) and, strictly speaking, the average duration of the interval between successive bouts of the behaviour (which should also be long relative to the sample interval). Of course, if a very short sample interval is used then the practical benefits of time sampling are lost, in which case continuous recording might as well be used instead.

A number of empirical studies have shown that instantaneous sampling can give a good approximation to the proportion of time spent performing a behaviour pattern (e.g., Dunbar, 1976; Leger, 1977; Simpson & Simpson, 1977; Rhine & Flanigon, 1978; Tyler, 1979). Choosing the right length of sample interval is discussed in 4.F below.

2 *Applications.* Instantaneous sampling is used for recording behavioural states that can unequivocally be said to be occurring or not occurring at any instant in time; for example, measures of body posture, orientation, proximity, body contact, or general locomotor activity. Instantaneous sampling is *not* suitable for recording discrete events of short duration. Neither is it suitable for recording rare behaviour patterns, since a rare behaviour pattern is unlikely to be occurring at the instant of any one sample point and therefore will usually be missed.

One potential source of bias with instantaneous sampling is the observer's natural tendency to record conspicuous behaviour patterns even if they occur slightly before or after the sample point. The sample point is therefore stretched out from an instant to become a window of finite duration, making the sampling no longer 'instantaneous'. If, as is likely, this is mainly done with the more noticeable or important behaviour patterns, then these will tend to be over-estimated relative to less prominent behaviour patterns.

Note that instantaneous sampling is sometimes confusingly referred to as 'scan sampling' in the behavioural literature.

E One-zero sampling

In one-zero sampling, as with instantaneous sampling, the recording session is divided up into short sample intervals. *On the instant of each sample point (i.e., on the 'beep'), the observer records whether or not the behaviour pattern has occurred during the preceding sample interval.* This is done irrespective of how often (or for how long) the behaviour pattern has occurred during that sample interval (see Fig. 4.4). An equivalent procedure is to record the behaviour pattern when it first occurs, rather then waiting until the end of the sample interval.

1. *Measure obtained.* The score is expressed as *the proportion of all sample intervals during which the behaviour pattern occurred.* For example, if a behaviour pattern occurred during 50 out of the 120 15-s sample intervals in a 30-min recording session, the score would be $50/120 = 0.42$. Note that, as with instantaneous sampling, one-zero sampling gives a single, dimensionless score for the whole recording session. Again, individual sample points within a recording session obviously cannot be treated as statistically independent measurements.

Fig. 4.4. One-zero sampling. The behaviour pattern, denoted here by the black bars, is scored according to whether or not it has occurred during the preceding sample interval.

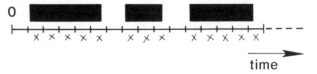

time

One-zero sampling does not give true or unbiased estimates of durations or frequencies. The proportion of sample intervals in which the behaviour occurred to any extent cannot be equated either with the length of time spent performing the behaviour, or with the number of times the behaviour occurred. This point must be emphasised even more strongly than in the case of instantaneous sampling since, in addition to not necessarily being accurate, one-zero sampling also introduces bias. One-zero sampling consistently *over*-estimates duration to some extent, because the behaviour is recorded as though it occurred throughout the sample interval when it need not have done so in reality. One-zero sampling also tends to *under*-estimate the number of bouts performed, because the behaviour could have occurred more than once during a sample interval. The shorter the sample interval relative to the average duration of the behaviour pattern, the more closely one-zero sampling approximates to instantaneous sampling. Chow & Rosenblum (1977) present methods for estimating frequencies and durations from time-sampled records.

If one-zero scores are compared – either between subjects or for different occasions – then problems can arise unless the mean bout length of the behaviour remains roughly constant. This is because the error in estimating frequency or duration depends on the ratio of mean bout length to sample interval. Thus, if the mean bout length of the behaviour varies between individuals (or, for the same individual, varies between different recording sessions), then the error in estimating frequency or duration will also vary (Dunbar, 1976).

Scores obtained using one-zero sampling are sometimes referred to as 'Hansen frequencies', but this term is best avoided because it encourages the mistaken view that one-zero scores actually are frequencies. One-zero sampling is also sometimes referred to as 'fixed-interval time span sampling'.

2 *Applications.* Because of the problems outlined above, one-zero sampling should be used with caution. Indeed, some authorities assert that one-zero sampling should *never* be used. Our own view is that one-zero sampling is valuable for recording certain types of behaviour patterns, for which neither continuous recording nor instantaneous sampling are suitable. This issue is discussed in 4.G.

F Choosing the sample interval

The size of sample interval used in time sampling will depend on how many categories are being recorded, as well as the nature of the behaviour. The shorter the sample interval the more accurate a time-sampled record will be. However, the shorter the sample interval the more difficult it is reliably to record several categories of behaviour at once – especially if the behaviour is complicated or occurs rapidly.

In practice, it is always necessary to balance the theoretical accuracy of measurement (which requires the shortest possible sample interval) against ease and reliability of measurement (which requires an adequately long interval). If the sample interval is too short, observer errors can make recording less reliable than if a slightly longer interval had been chosen. Thus, the sample interval should be the shortest possible interval that allows the observer to record reliably, after a reasonable amount of practice.

The best sample interval depends on what is being measured and is partly a matter of trial and error. To give some idea, though, many observers use a sample interval in the range from 10 s to 1 min, with sample intervals of 15, 20 or 30 s being most common under laboratory conditions. Field studies, especially those in which long recording sessions are used, tend to employ longer sample intervals.

Rather than relying on a combination of common sense and trial-and-error the sample interval can be chosen objectively, although this requires a considerable amount of additional work. First, a fairly large sample of the behaviour must be measured using continuous recording, in order to give a true picture of what actually happened. Scores are then calculated for each category as though the behaviour had been recorded using time sampling with various sample intervals (e.g.,10, 20, 30,...s). The discrepancy between the continuous record and the simulated time sampling measure can then be calculated for each sample interval. The error due to time sampling will increase as the simulated sample interval becomes larger. However, it may be possible to distinguish an obvious 'break-point', above which time sampling is too inaccurate, but below which it gives a reasonable approximation to continuous recording. This point marks the longest sample interval that can be used if the record is to be reasonably accurate for that category of behaviour.

In most studies, of course, several behavioural categories are

Fig. 4.5. One way of choosing the length of sample interval used in time sampling. First, a pilot sample of behaviour is measured using continuous recording (CR). This provides a true score for each category of behaviour. The score for each category is then calculated as though it had been recorded using time sampling with various different sample intervals (in this example, 10, 20, 30, 60 and 120 s). The percentage difference between the true (CR) score and the time-sampled score is then calculated for each sample interval, for each category of behaviour. The histogram shows the proportion of categories where the discrepancy between the continuous and time-sampled scores is 10% or less. In this example, a sample interval of 20 s would provide a good approximation to continuous recording, whereas longer sample intervals would introduce substantial inaccuracies for several categories (after Francis, 1966). Note that a different criterion (say, 5% or 15%) might be used to define the maximum acceptable discrepancy between the continuous and time-sampled scores.

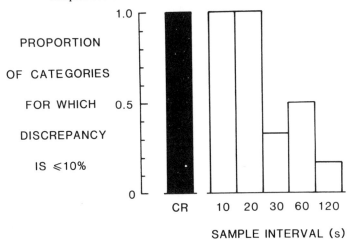

recorded and the sample interval used must be suitable for all of them. The simplest approach here is to specify the maximum acceptable discrepancy between the continuous record and the time-sampled measure for any category (say, 10%), and plot the proportion of categories where this condition is satisfied as a function of the sample interval. An example is shown in Fig. 4.5.

One problem with this whole process is the need initially to measure the behaviour using continuous recording in order to provide a true record for comparison. In many cases time sampling is used precisely because continuous recording is not practicable, which would obviously rule out this procedure.

G The advantages and problems of time sampling

As we have pointed out, time sampling methods are not perfect. Neither instantaneous sampling nor one-zero sampling give accurate estimates of frequency or duration unless the sample interval is short (see Fig. 4.6). Furthermore, time sampling is not generally suitable for recording sequences of behaviour unless the sample interval is very short (see 4.H). This is because with one-zero sampling two or more behaviour patterns can occur within the same sample interval, while instantaneous sampling can miss changes in behaviour which occur between sample points.

However, several empirical studies have shown that time sampling can in practice give quite good estimates of durations or frequencies. For example, by re-analysing data as though they had been acquired using one-zero and instantaneous sampling, Tyler (1979) found that both methods give accurate approximations to continuous recording, depending on the particular characteristics of the behaviour. One-zero sampling is usually less satisfactory than instantaneous sampling, but actually produces better estimates of frequencies and durations for certain types of behaviour.

A major practical advantage of time sampling is that, by condensing the information recorded and thereby reducing the observer's work load, it allows a larger number of categories to be measured than is possible with continuous recording. This can be an important consideration, especially in a preliminary study where it may be necessary to record a large number of categories. Time sampling also allows a larger number of subjects to be studied, if the individuals in a group are watched cyclically (i.e., using a scan sampling procedure; see 4.A.3). For example, to record the behaviour of a group of 12 subjects the observer might look at them cyclically, noting the behaviour of each subject (using instantaneous sampling) every 15 s, thereby watching each subject once every 3 min.

Fig. 4.6. Comparison between *continuous recording* (CR) and the two types of time sampling: *instantaneous sampling* (IS) and *one-zero* sampling (1/0). The black bars on the upper trace represent four successive occurrences of a behaviour pattern (B), during an observation period lasting t units of time and divided into 16 sample intervals. Assuming (arbitrarily) that each sample interval lasted 10 units of time (i.e., $t = 160$ units of time), then the following scores would be given by the various recording rules:

Continuous recording: Total duration $= a + b + c + d = 39 + 16 + 12 + 28 = 95$ units of time. Mean duration $= 95/4 = 23.8$ units of time. Proportion of time spent performing the behaviour pattern $= 95/160 = 0.59$. Frequency $= 4/t = 0.025$ per unit time. Total number of occurrences $= 4$.

Instantaneous sampling: Score $= 9/16 = 0.56$.

One-zero sampling: Score $= 13/16 = 0.81$.

Note that instantaneous sampling gives a reasonably close approximation to the proportion of time spent performing the behaviour pattern (0.56 versus 0.59) and accurately records that four separate 'bouts' occurred. One-zero sampling considerably over-estimates the proportion of time spent (0.81 versus 0.59) and records only three separate 'bouts', rather than four.

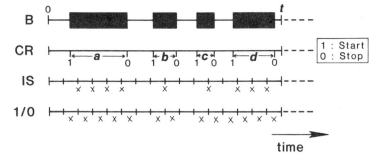

Because it is simpler and less demanding than continuous recording, time sampling also tends to be highly reliable (e.g., Rhine & Linville, 1980). Furthermore, some types of behaviour occur too rapidly for each occurrence to be recorded, making time sampling a necessity. The practical benefits of time sampling are, of course, achieved at the expense of preserving less information about the behaviour than does continuous recording.

The fact that one-zero sampling does not give unbiased estimates of frequency or duration has led some authorities, for example Altmann (1974), to argue that it should never be used. We disagree with this view, and not just because one-zero sampling is easy and reliable to use.

One-zero sampling is, arguably, the only practicable method for recording a type of behaviour pattern encountered in many studies, namely *intermittent* behaviour. Here we are referring to behaviour patterns which start and stop repeatedly and rapidly, and last only briefly on each occasion; for example, play behaviour in some species. In such cases, continuous recording or instantaneous sampling are not practicable, since it is difficult (or impossible) to record each occurrence of the behaviour, or to specify at any one instant whether or not the behaviour is occurring. However, it is usually possible to state unequivocally whether or not the behaviour has occurred during the preceding sample interval. To give a specific example, in a study of social play in captive rhesus monkeys, White (1977) found that one-zero sampling was the only effective way of recording some categories of play.

A second point is that one-zero scores are, arguably, valid measures of behaviour in their own right, in so far as they provide a meaningful index of the 'amount' of behaviour. One-zero scores are often highly correlated with both frequency and duration measures of the same behaviour (Leger, 1977; Rhine & Linville, 1980), which means that they give a composite measure of 'amount' of behaviour (see Kraemer, 1979*a*). In contrast, frequency and duration measures of the same behaviour are not always highly correlated with one another. In addition, we see no compelling reason why behaviour patterns should necessarily be described only in terms of frequency or duration (see Rhine & Linville, 1980; Smith, 1985). Indeed, in some cases a one-zero score may actually be a more realistic index than frequency or duration (Slater, 1978; Smith, 1985).

A final point worth emphasising is that instantaneous sampling, one-zero sampling and continuous recording can all be used simultaneously for recording different categories of behaviour (see also 5.B on check sheets).

H Recording and analysing sequences

1 *Recording sequences.* For some purposes, the object of interest is the sequence of behaviour, not the frequencies or durations of the component behaviour patterns. For example, an investigator may wish to record the sequence of display movements in the courtship behaviour of a bird. If the various behaviour patterns in a sequence are mutually exclusive (that is, only one can occur at any one time), then recording the sequence simply involves noting down each occurrence. For example, with three different behaviour patterns (A, B and C), the record might be of the form: ABAAACBA-BAAACB... When recording only sequence information, the observer would normally start to record the behaviour when the sequence started, rather than at a predetermined time as is generally the case when recording frequencies or durations.

If time sampling methods are used (4.B), information about the sequence of events within each sample interval is generally lost – although some sequence information is preserved on a crude time scale, accurate to the nearest sample interval. However, we do not recommend time sampling for recording sequences since it is possible to miss some types of behaviour altogether using these methods (see 4.G).

2 *Analysing sequences.* A sequence in which the component behaviour patterns always occur in the same order is described as **deterministic.** In reality, sequences of behaviour are usually variable, but nonetheless exhibit some degree of predictability. These are referred to as **stochastic** or **probabilistic** sequences. Sequences which show no temporal structure, where the component behaviour patterns are sequentially independent of one another, are referred to as **random** sequences. In a random sequence, one behaviour pattern can be followed by any other (including itself) with equal probability. The conditional probability that one behaviour pattern follows another – that is, the probability that B follows A, given that A has occurred, P (B|A) – is referred to as a **transition probability.**

Markov analysis is a method for distinguishing whether a sequence of behaviour is random or whether it contains some degree of temporal order. A **first-order** Markov process is one in which the probability of occurrence for the next event depends only on the immediately preceding event. If the probability depends on the two preceding events then the process is described as **second-order**, and so on. Sequences are analysed by comparing the actual number of times each transition occurs with the number of such transitions that would be expected if the sequence were random.

To take a highly simplified example, suppose only two different behaviour patterns (A and B) can occur. Four types of transition are therefore possible: AA, AB, BA or BB. If the sequence is random, and if A and B occur equally often on average, then each type of transition should occur equally often (i.e., with probability 0.25). To test this, a **transition matrix** is constructed, showing the actual transition probabilities. Suppose, for example, that B frequently follows A and vice versa (i.e., the transition probabilities of AB and BA are high) and there are few repeats (i.e., the transition probabilities of AA and BB are low). In this case, a transition matrix of the type shown in Fig. 4.7 might be obtained. (A real matrix would generally have more than four cells; an example of a larger (5 × 5) transition matrix is given by Slater, 1983, p. 21). If the actual transition probabilities differ significantly from the chance level (in this case, 0.25), then the behaviour patterns are not sequentially independent. A chi square test is sometimes used to calculate whether, overall, the observed transition probabilities depart significantly from the random model, although the assumptions on which this test is based are often violated by behavioural data (see Slater, 1973; Fagen & Mankovich, 1980).

If a sequence is scored in such a way that behaviour patterns can repeat themselves (as, for example, in the sequence ABBBAAB...), then problems can arise in deciding whether a given behaviour pattern has stopped and started again, or whether the same act is merely continuing (for example, BB versus B). The way in which this problem is handled greatly affects the conclusions drawn from a transition matrix. One solution is simply not to consider repetitions, but to deal only with transitions between different behaviour patterns (Slater, 1973), although this is, of course, only helpful when dealing with three or more behaviour patterns.

Transition matrix analysis assumes that the transition probabilities

Fig. 4.7. A highly simplified example of a transition matrix for analysing the sequence shown above it, which comprises only two different behaviour patterns (A and B). The matrix shows the empirical transition probabilities for the four different types of transition (A|A, B|A, A|B, B|B). For example, the bottom left-hand cell shows that the conditional probability of B given that A has occurred (B|A) is 0.45 (9 out of 20 transitions). For comparison, the transition probabilities under a random model (0.25 for each type in this example) are shown in parentheses. The matrix confirms that A and B tend to alternate (i.e., the probabilities of B|A and A|B are high) while repeats are rare (i.e., the probabilities of B|B and A|A are low).

sequence: ABABABBABABAABABABABA

1st behaviour pattern

		A	B
2nd behaviour pattern	A	0.05 (0.25)	0.45 (0.25)
	B	0.45 (0.25)	0.05 (0.25)

remain constant across time (the assumption of **stationarity**). However, this assumption is often likely to be violated – especially in long sequences of behaviour or sequences describing interactions between two or more individuals. The internal consistency of sequential data can be tested by analysing separately data from different parts of the record.

I Rhythms

Rhythmic variations in behaviour are found in many species, with periods ranging from a few minutes to several years.

1 *Terminology.* If a behaviour pattern is repeated over time, such that the distribution of intervals between successive occurrences is roughly regular (rather than random), or if the rate of occurrence varies in a roughly cyclical manner, then the behaviour is said to be **rhythmic**. If the intervals between successive occurrences are known to be equal (within specified limits of variation), then the behaviour is said to be **periodic** (or cyclic). The term periodic is more restrictive than rhythmic and is applied only when the regularity of variation has been demonstrated. In mathematical terms, a time-dependent variable, $x = f(t)$, is said to be periodic with period T if, for all values of time, $f(t) = f(t + T)$. A simple example of a periodic function is a sine wave.

Data describing a series of events as a function of time, such as regular occurrences of a behaviour pattern, are known as a **time series**. Periodic variations over time are conventionally described using the same terms as are applied to vibrations or waves. The **period** (or wavelength) of a rhythm is the interval between successive peaks (or troughs) on the wave. The **amplitude** is the magnitude of change between peaks and troughs; that is, the maximum range of variation in behaviour.

The most familiar type of behavioural rhythm is the **circadian rhythm**, with a period of approximately 24 h. However, rhythmic variations over different time scales are also found, including **ultradian rhythms** (periods considerably less than 24 h), **infradian rhythms** (periods considerably more than 24 h, such as lunar cycles) and **circannual rhythms** (periods of approximately one year, such as annual migration or hibernation). If the characteristics of a time series (period, amplitude, etc.) do not change over time, the time series is said to be **stationary** (or steady-state).

Many behaviour patterns exhibit complex rather than simple rhythms, with several rhythms of different periods superimposed on one another. (These correspond to the harmonics superimposed on the fundamental oscillation of a vibrating string.) For example, one or more ultradian rhythms in activity are sometimes superimposed on a dominant circadian rhythm (e.g., Broom, 1980; Kavaliers, 1980).

2 *Methods for detecting rhythms.* Rhythmic variation in behaviour can be detected using four principal methods of analysis.

(*a*) *Plotting the data.* The behavioural variable, such as frequency or duration, is plotted on the *y*-axis as a function of time. A periodicity of large amplitude should, if present, be fairly obvious. For example, Broom (1980) measured the locomotor activity of isolated domestic chicks in successive 30-min periods over several days after hatching. Simply plotting activity against time on a graph showed the presence of an obvious circadian rhythm.

(*b*) *Autocorrelation.* If a variable oscillates with period T, there will be a maximum positive correlation between the values at times T, $2T$, $3T$,..., nT. Autocorrelation involves calculating all possible correlations between values of the variable at different intervals and is an effective way of detecting the dominant rhythm in a time series.

(*c*) *Spectral analysis.* This works on the principle that any complex rhythm can be analysed as the sum of simple rhythms of different wavelengths. A commonly used form of spectral analysis is **Fourier analysis**, in which a complex rhythm is analysed as the sum of an infinite number of sinusoidal oscillations of different wavelengths. A **spectrogram** (or spectral density plot) is a plot of how oscillations of different wavelengths contribute to the total variance of the time series. Spectral analysis is particularly effective at analysing complex time series composed of multiple rhythms of different wavelengths. For example, it can detect the presence of one or more ultradian rhythms superimposed over a dominant circadian rhythm (e.g., Campbell & Shipp, 1974; Broom, 1980). Another common application of spectral analysis is in decomposing the complex brain-wave patterns of an electroencephalogram (EEG) into component waves of different frequencies.

(*d*) *Multiple regression analysis*. This is another way of calculating how rhythms of different wavelengths contribute to the total variance in the data. In addition, multiple regression allows simple trends in the data resulting from non-stationarity (such as linear changes in amplitude over time) to be removed. This method, in combination with spectral analysis, is effective at analysing complex rhythms containing multiple periodicities and trends (Box & Jenkins, 1970). For example, Broom (1980) analysed the activity patterns of domestic chicks using autocorrelation, spectral analysis and multiple regression analysis and detected the presence of various ultradian rhythms in activity (with periods of 1.5–4 h and 30 min) superimposed over a dominant circadian (24-h) periodicity.

Bear in mind that even if spectral analysis or multiple regression analysis does indicate the presence of a periodicity in behavioural data, the amplitude of the variation might be small and the rhythm may have little or no biological significance.

J Further reading

Altmann's (1974) review of observational recording methods has been influential, particularly in discouraging the use of one-zero sampling. See also Dunbar (1976), Slater (1978) and Kraemer (1979*a*). A different view of one-zero sampling is given by Tyler (1979) and Smith (1985), who defend its use on practical and theoretical grounds. Others have also concluded, for a variety of reasons, that one-zero sampling can be an acceptable method for recording certain types of behaviour (e.g., Rhine & Flanigon, 1978; Slater, 1978; Rhine & Linville, 1980; Smith & Connolly, 1980). A comprehensive explanation of recording methods is given by Lehner (1979).

Analysing sequences of behaviour is a large and complicated topic. For an introduction to the many issues involved, see Hutt & Hutt

(1970, ch.9), Lehner (1979, pp. 269–278) and Dawkins (1983). Slater (1973) gives a good review of sequences. More advanced accounts of methods for analysing sequences are given by Chatfield & Lemon (1970), Andersson (1974), Morgan *et al.* (1976), Bekoff (1977), Cane (1978), Fagen & Young (1978) and Douglas & Tweed (1979). Slater (1983) discusses sequences of behaviour in the specific context of communication.

For methods of detecting and analysing temporal rhythms in behavioural data, see the books on time series analysis by Box & Jenkins (1970) or Gottman (1981), or the papers by Fagen & Young (1978), Broom (1979) or Kraemer *et al.* (1984). Specific examples of behavioural rhythms and methods used to analyse them are given by Campbell & Shipp (1974), Broom (1980), Kavaliers (1980) and Regal & Connolly (1980). The biological bases and significance of rhythms are discussed by McFarland (1981, 1985) and Huntingford (1984*a*, pp. 160–166).

5

The recording medium

A The options available

The choice of the medium, or physical means, used to record behavioural observations has important effects on the sorts of data that can be collected and the sampling techniques that can be used. Six basic methods of recording behaviour are available: film or videotape; written or dictated verbal descriptions; pen recorders; automatic recording devices; check sheets; and computer event recorders. In most quantitative studies, observations are recorded either on check sheets or on the keyboard of a computer event recorder, so these methods are described separately in 5.B and 5.C. Using check sheets or an event recorder presupposes that the observer has formulated a set of discrete behavioural categories (see 3.C).

1 *Film or videotape* gives an exact visual record of the behaviour, which can subsequently be slowed down for analysis. This is useful for studying behaviour that is too fast or too complex to analyse in 'real time'. Similarly, exact records of vocalisations can be made with an audio tape-recorder and the sound patterns analysed later using instruments such as a sound spectrogram. The major advantage of recording behaviour in this way is that the record can be analysed repeatedly and in different ways.

However, filming or videotaping is rarely a complete replacement for observation. At some stage it is always necessary to **code** the behaviour – that is, transcribe the record into quantitative measurements relating to specific behavioural categories, by recording on a check sheet or event recorder whilst watching the film or videotape. Clearly, it is far more efficient (if possible) to code the behaviour as it actually happens and not postpone the measurement process by filming or videotaping it instead. Behaviour is usually easier to observe and analyse 'live' and in context, rather than by watching it later on a small screen. Thus, film or videotape should normally be used only if there is some reason to do so – for example, if the behaviour is very fast or complicated and therefore requires repeated or very detailed analysis. If the behaviour is rare or of particular interest, film or videotape is a useful back-up to live observation, ensuring that nothing is lost.

A major problem with film or videotape analysis is that it can be exceedingly time-consuming: behaviour that perhaps lasted only a few minutes may take hours to analyse. Furthermore, the ease with which a film or videotape can be replayed may lead to a temptation to analyse the record repeatedly and in ever greater detail, whereas it might be better to use this time for collecting more data. Another possible drawback of film or videotape is the restricted field of view and depth of focus.

2 *A verbal description* of the behaviour can be recorded in the form of longhand written notes or dictated into a miniature tape-recorder. (A tape-recorder is particularly convenient when recording outdoors in bad weather.) Verbal descriptions are valuable during informal pilot observations and for recording rare events which are not a central feature of the study. In addition, dictating a verbal description is sometimes necessary for recording complex behaviour involving several categories, which cannot reliably be coded directly onto a check sheet or keyboard.

Spoken or written records must at some stage be coded if they are to be quantified and, as with film or videotape, transcription can be very time-consuming. Some researchers report that transcribing verbal records may take 10 to 15 times as long as the original observation time (e.g., Holm, 1978).

3 *Pen recorders* (or chart recorders) were once popular for recording behaviour, but have now largely been superseded by computer event recorders (see 5.C). A pen recorder is an electromechanical device similar to a polygraph, in which a strip of paper moves at a constant speed beneath several pens. Each pen makes a continuous line, or trace, on the paper and is deflected by closing a switch. The behavioural record therefore consists of several traces, each representing a single category of behaviour, the deflections on each trace denoting individual events or states.

The major disadvantage of pen recorders is the need to transcribe the chart record into numerical form, which requires laboriously counting or measuring with a ruler every deflection on each trace (see Hutt & Hutt, 1970, ch. 5).

4 *Automatic recording devices* can be used to measure some kinds of behaviour. It is relatively simple to provide an electrical signal indicating when a lever is pressed, when an animal stands on a particular area of the floor, when it vocalises, or even when it moves. An enormous range of devices is available, many giving a record suitable for storage and analysis on a computer. The archetypal example is the **Skinner box** used by experimental psychologists for studying operant conditioning. **Activity monitors** can record movements automatically by means of infra-red, ultrasonic, capacitative or microwave detectors, the interruption of optical or infra-red light beams, or the disturbance of sensitive microswitches. Physiological measures, such as heart rate, rectal temperature, blood pressure, electromyogram (EMG), or electroencephalogram (EEG) can be recorded remotely from sensors or electrodes using biotelemetry techniques.

A computer can be used both to run an experiment (for example, by turning equipment on and off, administering rewards, or opening and closing food hoppers) and to record the results (for example, by counting lever-presses in a Skinner box or revolutions of a running-wheel, or monitoring data from physiological sensors).

Signals from measurement devices can be recorded by a computer, and the availability of powerful and inexpensive microcomputers makes this an attractive way of recording behaviour (see 5.C.6). Sensors and other measurement devices can be connected to a computer using an **analogue-to-digital converter**, a wide variety of which are now available for use with standard mini- and micro-computers. The list of possibilities is enormous and is limited mainly by the ingenuity of the experimenter. Our primary concern in this book, however, is with observational measurement techniques. For more extensive accounts of automatic measurement and biotelemetry techniques, see Amlaner & Macdonald (1980), Amlaner (1981) and Mech (1983).

Finally, we should stress that even in the most exquisitely and comprehensively automated study, in which virtually everything has been 'wired up', it is still essential to *watch* what is going on.

B Check sheets

A check sheet is a simple paper-and-pencil tool which provides a cheap, flexible and surprisingly powerful way of recording observations. With practice and a correctly designed check sheet, a considerable amount of information can be recorded reliably and with reasonable accuracy.

1 *The basic design* of a check sheet is a grid, with *columns* denoting different categories of behaviour and *rows* denoting successive sample intervals (Fig. 5.1). At the end of each sample interval, usually indicated by an audio beeper (Appendix 2), the observer starts recording on the next row of the check sheet. The choice of sample interval (the time represented by each successive row on the check sheet) has been discussed in 4.F. Columns denoting related categories of behaviour, or behaviour patterns that tend to occur together, should be grouped on the check sheet. It is always a good idea to leave one or two blank columns for additional information or new categories that may be added during preliminary observations. Another helpful feature is a remarks column, which can be used to note down unexpected events or incidental information that may later be useful in interpreting the results.

Some field studies combine focal, scan and *ad libitum* sampling (see 4.A), with different parts of the same check sheet set aside for categories recorded using the different sampling rules. In addition, all three recording rules described in 4.B (continuous recording, one-zero sampling and instantaneous sampling) can be used on the same check sheet, for different categories of behaviour.

With instantaneous sampling, behaviour is recorded by marking 'on the line' at each sample point. With one-zero sampling, a mark is made 'in the box' when the behaviour pattern first occurs. For continuous recording of frequencies, each occurrence is marked down within the appropriate sample interval. The exact frequencies of events as well as their sequence can be recorded relatively easily on a check sheet.

Recording precise durations on a check sheet is difficult, since the actual times at which the behaviour started and stopped must be written down from a digital stopwatch (although skilled observers can learn to do this quickly and accurately). A cruder way of recording durations is simply to mark within the appropriate sample interval (the current row) whenever the behaviour stops and starts.

Fig. 5.1. The basic design of a check sheet. Columns denote different categories of behaviour. Rows denote successive sample intervals.

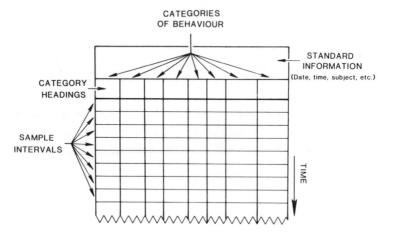

This is obviously easier and quicker than reading a stopwatch and writing down the actual times, but timing is only accurate to the nearest sample interval. If the sample interval is very short relative to the average bout length, however, this may be adequate for many purposes.

An experienced observer can learn to record behaviour without looking down at the check sheet very often. Some observers (e.g., Scott, 1970) have found that placing a wire grid over the check sheet allows the pencil to be guided to the right position by touch, obviating the need to look down. One hand is used to locate the correct square on the grid, the other hand to write.

2 *The number of columns* on a check sheet will obviously depend on the number of categories recorded, but can be reduced by using different symbols within each column to denote different categories or sub-categories of behaviour. For example, within a single column labelled 'vocalisations', different symbols can be used to denote particular types of vocalisation. Symbols can also be used to identify individual animals; for example, the initiator and recipient of a social interaction. If too many columns are used, recording will become difficult and therefore inaccurate.

If data from check sheets are eventually to be analysed using a computer, then the design of the check sheet should take this into account, preferably by allowing the results to be transcribed directly onto a computer keyboard in a suitable format. A check sheet design that necessitates a two-stage transfer of data, from check sheet to a coding sheet and thence onto a computer, wastes time and creates additional opportunities for error. Bear in mind, though, that data should be easily accessible before they are analysed (see 5.C and 9.B). If the computer does not allow the data to be summarised and subjected to exploratory analysis, then it is probably best to transcribe the data onto summary sheets before putting them onto a computer.

An example of a fairly typical, though simple, check sheet is shown in Fig. 5.2.

C **Computer event recorders**

1 *Basic principles.* Computer event recorders are devices for recording observations directly onto a keyboard and storing them in a digital form, suitable for analysis by a computer. Originally, event recorders used custom-built keyboards and circuitry to encode observations onto paper punch-tape or magnetic tape for subsequent storage and analysis on a main-frame computer (e.g., Tobach *et al.*, 1962; Dawkins, 1971; White, 1971; Sackett *et al.*, 1973). With the advent of cheap, portable computers, however, observations can now be recorded directly onto the keyboard of a microcomputer (e.g.,

Fig. 5.2. A simplified example of a check sheet, in this case designed for recording the behaviour of a rhesus monkey mother and her infant. Five categories of behaviour were recorded, using continuous recording (CR), instantaneous sampling (IS) or one-zero sampling (1/0): *Mother and infant in ventro-ventral contact* (IS); *Groom* (IS); *Approach* (CR); *Leave* (CR); and *Eat* (1/0). In the example shown, the mother and infant were initially in ventro-ventral contact. The mother (M) then left the infant. The infant (I) then approached the mother and the mother groomed the infant. Finally, the infant left the mother and started to eat. For clarity, only the top part of the check sheet is shown.

MOTHER: Gladys	DATE: 18 Aug. 85		PAGE: 1
INFANT: Mabel	START TIME: 14·30	OBSERVATION	
GROUP: B	OBSERVER: PHM	SESSION No.: 3	
	TEMP: 19° C		

VENTRO-VENTRAL CONTACT	GROOM	APPROACH	LEAVE	EAT	REMARKS
✓					
✓					
✓					
			M		
	M	I			
	M				Grooming belly
	M				
			I	I	
				I	

Flowers, 1982; see Fig. 5.3). All that is needed nowadays for many applications is an ordinary microcomputer and suitable event-recording software for encoding and interpreting the key-presses. (Incidentally, to use the terminology of 3.F, most 'event recorders' are actually capable of recording both *events* and *states*.)

Many different sorts of hardware can be used, but all event recorders have certain features in common. Observations are recorded as key-presses, either on the conventional 'QWERTY' keyboard of a microcomputer or on a custom-built keyboard. Each key denotes a particular category of behaviour or a particular subject. Every time a key is pressed the identity of the key, together with the time at which it was pressed (from an internal clock), are recorded. At the end of an observation session the observer can usually obtain a 'hard copy' print-out of the data, transfer the data to a larger computer for storage and analysis, or both.

An alternative way of using a computer event recorder is as a check sheet emulator. In this case, information is recorded at regular intervals (sample intervals) in the form of a string of key-presses, the final key-press of each string denoting the timing. The advantage of this method is that complex behaviour, such as social interactions, can be recorded using only simple event-recording software. This gives the event recorder the advantages of a check sheet (notably flexibility and simplicity) in addition to allowing accurate timing and obviating the need to transcribe data manually at a later stage.

2 *The pros and cons of using event recorders.* Computer event recorders offer three major advantages over check sheets:

(*a*) Durations are recorded with considerably better precision than is possible with a check sheet.

(*b*) The observer can record more rapidly-occurring streams of behaviour and use more categories than is practicable with a check sheet.

(*c*) They eliminate the laborious and error-prone task of transcribing raw data from check sheets into numerical form, which can sometimes take longer than collecting the data in the first place.

However, event recorders also have certain perils and problems attached to their use:

(*a*) They offer a degree of sophistication that is unnecessary for many jobs. There is no point in using a complex and expensive piece of machinery if paper and pencil would do just as well. As a general rule, it is wise to choose the simplest recording technique that will do the job.

(*b*) If an event recorder is not highly dependable – and by no means all are – both time and irreplaceable data can be lost.

(*c*) A related point is that although they eliminate the need to transcribe check sheets, event recorders can introduce their own form of manual labour, namely the need to check for recording errors. If the machine is not completely trustworthy, checking every record for errors can take as long as transcribing check sheets.

(*d*) Event recorders, especially the simpler sorts, are not always as adaptable as check sheets, which can be designed specifically to suit each application. The recording medium should be designed around the behavioural categories to be recorded, and not vice versa.

(*e*) Event recorders can erect an undesirable barrier between the observer and the data, particularly if the data have to be transferred to another computer before the observer can print them out. To get the most out of a study it is essential to become familiar with the data as they are acquired – in other words, to have the numbers on the desk in front of you and literally *look* at them. Event recorders tend to encourage the habit of siphoning data directly into computerised 'black boxes' for statistical analysis without first inspecting the results. This practice can (and does) lead to blunders.

(*f*) The ease with which data can be collected using an event recorder can be a mixed blessing, since it is possible to collect too much information as well as too little. Using an event recorder should not be an excuse for being vague about the questions being asked or collecting irrelevant data.

(*g*) Preparing the software for an event recorder can present a major obstacle, especially to those who are not experts at writing computer programs (see 5.C.5). However, we expect general purpose event-recording programs to be increasingly available.

3 *The essential features for an event recorder.* Any event recorder, whether it is a custom-built machine or a standard microcomputer adapted for that purpose, should have the following features.

(*a*) The keyboard should have a sufficient number of keys to record all the categories, preferably without using multiple key-presses (such as SHIFTed keys). In the rare cases where the 50 or more keys of a conventional keyboard are too few for very complex behavioural observations, a custom-built keyboard with more keys may have to be used (see Fig. 5.3*b*)

(*b*) Data should be stored reliably, in a form that cannot accidentally be erased or lost and which is not subject to errors during transcription. Storage media that involve moving parts, such as magnetic tape or paper punch-tape, are not always perfect in this respect. Furthermore, transferring data to or from magnetic tape is slow compared with some other forms of data storage. In the more sophisticated machines, data are stored digitally in non-volatile solid state memory and transferred to a suitable long-term storage device, such as magnetic disc, at suitable intervals.

(*c*) Some form of visual feedback should be given to confirm which keys have been pressed. The display – which can be an alphanumeric liquid crystal display (LCD), light-emitting diodes (LEDs) or a cathode-ray visual display unit (VDU) – can also provide additional information, such as real or elapsed time, warning messages and an indication of key states ('on' versus 'off' for duration keys).

(*d*) Errors should be easily rectifiable by deleting mistaken key-presses.

(*e*) The machine should have sufficient memory to store as many key-presses as are likely to be needed in a single recording session. An adequate capacity for many purposes would be of the order of 4000 key-presses, allowing the user to record an average of one key-press per second for more than an hour.

(*f*) Perhaps the most important requirement of all is that an event recorder must be *dependable*. Few things are more wasteful and frustrating than losing time and data because of a faulty machine. The issue of dependability applies to software as well as hardware, since a reliable microcomputer can be turned into an unreliable event recorder by using programs containing errors. The quality and reliability of the hardware in most standard microcomputers are usually good. Moreover, mass-produced microcomputers are relatively cheap and it may cost considerably less to adapt a standard microcomputer than to design and build a specialised event recorder. (Most people these days would not attempt to build their own motor car, for example, unless they had very special requirements.) With standard hardware the user is not dependent on highly specialised personnel for maintaining the apparatus. Furthermore, a wide variety of 'add-on' equipment, such as analogue-to-digital interfaces and data storage devices, can be bought for many standard microcomputers.

The simplest type of event recorder is capable of recording frequencies or durations for several categories of behaviour, but may not be able to deal adequately with the complexities that arise when several subjects are watched simultaneously. For many purposes, it is necessary to record social interactions of the type 'Subject/Verb/Object'; for example, 'Animal A/approaches/Animal B', or 'Mother 2/nurses/Pups 1,3 & 5'. This type of event recording requires more sophisticated software for analysis (though not necessarily for recording) and, sometimes, a larger number of keys.

4 *The desirable features for an event recorder.* The following additional features are desirable, if not always necessary, in an event recorder.

(*a*) It should be possible for the user to 'customise' the keyboard by selecting keys to perform particular roles, allowing the machine to be used for more than one type of observation procedure. Some machines allow the user to assign each key to a particular role, usually by asking a series of questions each time the machine is switched on. Even in the simplest event recorder with fixed-function keys, the keyboard can be customised to some extent by placing labels on the keys.

(*b*) Most event recorders allow the user to transfer data to another computer for permanent storage and detailed analysis. However, a hard copy should also be available as soon as the recording session ends. This gives the observer immediate access to the results and also acts as a valuable insurance against losing data during the transfer. Ideally, the print-out should give both a complete record of each key-press and a summary in the form of overall frequencies and durations.

(*c*) Portability is an essential feature for many purposes, notably in field studies. It is now possible to buy powerful 'lap-top' portable microcomputers which are the size of a large notebook and can use the same software as conventional desk-top microcomputers (see Fig. 5.3*a*). These portable machines are battery-operated and use an LCD rather than a VDU but are, in most other respects, comparable to much larger machines.

(*d*) Whether or not portability is required, a battery back-up power supply is a sensible precaution for any event recorder. This is to prevent the machine from losing data if the mains electricity supply is interrupted. With a conventional desk-top microcomputer even a momentary interruption in the electricity supply will cause all data and software in the machine's central memory to be lost. Only data that have been transferred to non-volatile memory, such as magnetic disc or tape, are safe.

5 *Software for event recorders.* Computer processors operate using instructions written in a low-level language called **machine code**. To simplify the act of writing software, programmers use assemblers, compilers and interpreters. An **assembler** (such as the Z80 assembler in CP/M) is a program that generates machine code from a list of instructions written in mnemonic form. It simplifies the task of machine-code programming, but the instructions are incomprehensible to all but a few computer experts and the code will only run on a particular type of hardware.

A **compiler** (such as FORTRAN or PASCAL) interprets a listing that is more easily understood than an assembler listing, and turns it into machine code. The computer then runs the generated machine-code program, while the compiler and the original program listing are no longer required to be resident in the machine.

Fig. 5.3. (*a*) A commercially produced, portable microcomputer (in this example, an Epson PX-8) being used as a behavioural event recorder. The machine, which is one of several comparable machines now available, is small (29 × 22 × 5 cm), light (2.3 kg) and will operate for several hours from internal, rechargeable batteries. The visual display is an 80-column × 8-line LCD display. Fairly simple event-recording programs are written in a standard, high-level language (BASIC) and will run on different machines with relatively little modification. (Drawn from a photograph.) (*b*) An example of a custom-built computer event recorder (the Madingley 'MICRO'). Note the large number of keys provided. Sophisticated event-recording software is written in a specialised, low-level assembler language, requiring a particular type of hardware. (Drawn from a photograph.)

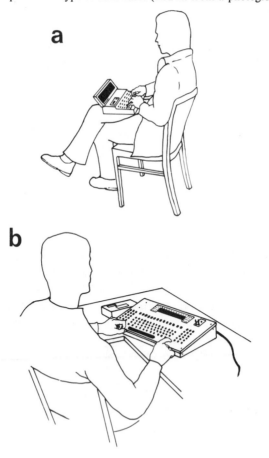

An **interpreter** (such as BASIC) is a program that is resident in the computer and interprets the listing, which is written in a sub-set of English, as it is running. Programs written in BASIC can be easily modified and adapted by the non-expert. Although there are many dialects of the language, conversion of a BASIC program to run on a different machine is usually straightforward. Thus, writing event-recording software in a high-level language such as BASIC has many advantages.

A computer running a BASIC program has to interpret the program as it goes along, whereas a compiled program has already been interpreted: the interpreter therefore runs more slowly. On some machines speed can become an important practical consideration and can place a limit on the complexity of some parts of the program. If the keyboard input routine were too slow, for example, it would limit the rate at which key-presses could be recorded. Furthermore, timing events with millisecond precision is generally not possible with high-level languages such as BASIC (Dlhopolsky, 1983), although this degree of precision is unnecessary for most types of behavioural observation – and would, anyway, exceed the observer's physical capacity. If very complex software is needed – for example, to record social interactions – then the only solution, other than using a faster computer, may be to write all or part of the program in machine-code or assembler.

6 *Automatic recording using a computer.* As we pointed out in 5.A.4, a computer can also be used to record data automatically from a switch or measurement device; for example, the lever in a Skinner box or an activity monitor. The distinctions made in 4.B between continuous recording and the two types of time sampling (instantaneous and one-zero) apply equally to automatic recording.

The computer equivalent of continuous recording is the use of an **interrupt routine:** a signal from an input line interrupts the computer's processor, which stops whatever it is doing at the time and records the occurrence of the input signal and the time when it happened. This technique is effective for recording the occurrence of behaviour patterns that can be indicated by a short pulse (the equivalent of *events*, in the terminology of 3.F), but is not appropriate for recording the durations of *states*, for which some kind of time sampling is required.

In the case of instantaneous sampling, the computer looks at its input channels at regular intervals and records whether or not the signal is present at that moment (the sample point). The problem with this technique is that short pulses will not generally occur at the same time as the sample point, and will therefore be missed. This can be overcome by artificially lengthening the signal pulses to something approaching the sample interval to ensure that most are counted. Alternatively, the equivalent of one-zero sampling may be used. To do this, the input pulse sets an external 'latch' circuit; the computer then records at regular intervals which latches are set and re-sets them (see Symonds & Unwin, 1982). In this case, multiple pulses arriving during the same sample interval would all be counted as one. If the sample interval is very short (say, 0.1 s), there is little to distinguish instantaneous sampling from one-zero sampling, and the resolution of both would approach that of continuous recording. Under these conditions, one-zero sampling using external latch circuits is probably the most versatile and simplest to implement. Programs for one-zero sampling can usually be written in a high-level language such as BASIC, whereas interrupt routines have to be written in machine code.

An important practical problem with time sampling using a computer is the vast number of data points that can be generated, many of which will contain no useful information. In view of this, it may be worth recording in memory only the presence of events (not their absence) and changes in states (not their continuation). This involves no loss of information, but can greatly reduce the number of data points that need to be stored in the computer's memory.

D Further reading

The applications of tape-recorders, film, videotape, check sheets, pen recorders and event recorders are reviewed by Hutt & Hutt (1970) and illustrated with examples from studies of various species. See also Holm (1978). Radio-tracking and telemetry are reviewed by Amlaner (1978, 1981), Macdonald (1978), Amlaner & Macdonald (1980) and Mech (1983). The design of check sheets is thoroughly covered by Hinde (1973); see also Hutt & Hutt (1970). The use of microcomputers as behavioural event recorders is discussed by Flowers & Leger (1982), while Kieras (1981) gives sound advice about possible pitfalls with computer equipment. Symonds & Unwin (1982) describe in simple terms how to use a microcomputer to record data automatically from several sensors, such as activity detectors.

6

The reliability and validity of measures

A Reliability versus validity

Measuring behaviour, like measuring anything else, can be done well or badly. When assessing how well behaviour is measured, two basic issues must be considered.

1 *Reliability* concerns the extent to which measurement is repeatable and consistent; that is, free from random errors. An unbiased measurement consists of two parts: a systematic component, representing the true value of the variable, and a random component due to imperfections in the measurement process (see 9.I). The smaller the error component, the more reliable the measurement.

Reliable measures, sometimes referred to as **good** measures, are those which measure a variable precisely and consistently. At least four related factors determine how 'good' a measure is:

(*a*) *Precision:* How free are measurements from random errors? This is denoted by the number of 'significant figures' in the measurement. Note that accuracy and precision are not synonymous: accuracy concerns systematic error (bias) and can therefore be regarded as an aspect of validity (see below). A clock may tell the time with great precision (to within a millisecond), yet be inaccurate because it is set to the wrong time.

(*b*) *Sensitivity:* Do small changes in the true value invariably lead to changes in the measured value?

(*c*) *Resolution:* What is the smallest change in the true value that can be detected?

(*d*) *Consistency:* Do repeated measurements of the same thing produce the same results?

2 *Validity* concerns the extent to which a measurement actually measures those features the investigator wishes to measure, and provides information that is relevant to the questions being asked. Validity refers to the relation between a variable (such as a measure of behaviour) and what it is supposed to measure or predict about the world.

Valid measures, sometimes referred to as **right** measures, are those which actually answer the questions being asked. To decide whether a measure is valid ('right'), at least two separate points must be considered:

(*a*) *Accuracy:* Is the measurement process unbiased, such that measured values correspond with the true values? Measurements are accurate if they are relatively free from *systematic* errors (whereas precise measurements are relatively free from *random* errors).

(*b*) *Specificity:* To what extent does the measure describe what it is supposed to describe, and nothing else?

For example, suppose an investigator wishes to assess how much milk a young mammal receives from its mother, and uses a behavioural measure – the total duration of suckling – to measure this. The behavioural measure is only valid for this purpose if there is a strong positive correlation between total duration of suckling and the amount of milk transferred. In some species, such as rats and pigs, the relation between suckling duration and milk intake is poor. Young piglets and rat pups spend a lot of time suckling, but obtain no milk for most of the time they suckle. Milk release only occurs during brief periods (less than a minute), which are separated by long intervals (20 min or more). Thus, suckling duration is not the *right* measure for assessing milk transfer in these species, even though it can be measured very precisely and consistently.

An example of how a behavioural measure can be validated is given by Ten Cate's (1985) work on sexual imprinting in zebra

finches. Studies of sexual imprinting have generally assumed that
the amount of singing a male bird directs towards a female during
short choice tests is a measure of his sexual preference for that
female, since singing is part of normal courtship. However, only
recently has this measure been directly validated by Ten Cate, who
showed empirically that the amount of singing was indeed highly
correlated with subsequent measures of sexual behaviour and pair-
formation.

Thus, it is relatively easy to devise behavioural categories, tests,
questionnaires or interview techniques that are believed to measure
some aspect of a subject's behaviour, personality or intelligence, but
it requires external evidence (validation) to demonstrate that they
actually measure what they are supposed to measure.

It is quite possible to obtain highly reliable (good) results using
biased, irrelevant or meaningless (wrong) measures. Methodological
rigour may sometimes have to be sacrificed in the interests of
measuring the things that really matter. If necessary, it is often
better to measure the right thing imperfectly rather than the wrong
thing extremely well.

In addition to the principal distinction between reliability and
validity ('good' versus 'right' measures), an important practical
consideration is **feasibility**. This concerns whether the proposed
measurement procedure is possible, practicable and worth while.
Does the information obtained justify the cost and effort required?
Assessments of feasibility should also include ethical considerations
(see 10.E).

B Intra-observer versus inter-observer reliability

Observers can be regarded as instruments for measuring
behaviour, in much the same way that, say, a voltmeter is used to
measure an electric potential. Just as measuring instruments can be
biased or imprecise, so errors in measuring behaviour can arise from
variation within or between observers. Two different measures of
reliability can be distinguished:

    1 **Intra-observer reliability** (or **observer consistency**) measures
       the extent to which a single observer obtains similar results

when measuring the same behaviour on different occasions (for example, when coding the same videotape twice).

2 **Inter-observer reliability** measures the extent to which two or more observers obtain similar results when measuring the same behaviour on the same occasion. This is a measure of agreement between different observers attempting to measure the same thing.

In any study involving two or more observers, two things must be verified: (*a*) that each observer consistently records in the same way on different occasions (i.e., that each observer exhibits good intra-observer reliability); and (*b*) that the observers are recording the same behaviour in the same way (i.e., that the inter-observer reliability for each category of behaviour is good). Obviously, the need to establish good intra-observer reliability also applies to studies involving one observer.

Even in studies involving only one observer, however, a demonstration of good *inter*-observer reliability is still valuable. This obviously requires using a second observer for some recording sessions. The point of measuring inter-observer reliability in a single-observer study is that the principal observer could be highly reliable at measuring the *wrong* behaviour (wrong, that is, with respect to the stated definitions). Good intra-observer reliability demonstrates internal consistency, but does not guarantee that another observer would record the same behaviour. Thus, even when all measurements are made by one observer, it is still helpful to demonstrate that a second observer could have produced a similar set of results using the same methods.

To assess intra-observer reliability, the observer measures the same sample of behaviour on two separate occasions. Typically, this involves videotaping or filming the behaviour. To assess inter-observer reliability, a sample of the behaviour is measured simultaneously by two (or more) observers, either 'live' or from a videotape or film recording. Both inter- and intra-observer reliabilities can be assessed by calculating the degree of association (correlation) or agreement (concordance) between the two sets of measurements.

C Measuring reliability using correlations

Reliability is often expressed as a correlation coefficient (see 9.G); either a Pearson (product moment) correlation coefficient (r) or a Spearman rank correlation coefficient (r_s). A correlation of +1.0 denotes a perfect positive association between two sets of measurements, while a correlation of zero denotes a complete absence of any linear association. A reliability correlation can be calculated for each measure or category of behaviour.

To measure intra-observer reliability, the observer codes each of n independent samples of behaviour (for example, n different videotapes) on each of two occasions. The reliability for each category of behaviour is then calculated as the correlation between the n pairs of scores. To measure inter-observer reliability, two observers simultaneously code each of n independent samples of behaviour (for example, during n different recording sessions). Again, the reliability of each category is calculated as the correlation between the n pairs of scores.

When quoting reliability, results should clearly state whether the correlations are Pearson or Spearman correlation coefficients and give the number of pairs of scores (n, the sample size) on which the correlations are based. Note that measuring reliability requires *independent* pairs of measurements (see 2.E). It is not correct to base a reliability correlation on a single sample of behaviour (for example, one observation session or videotape) which has been split up into several short sub-samples, since the measurements would not be independent.

Another point is that the test samples used to assess reliability should be fair (and preferably random) samples of the behaviour that is actually measured in the study. It would be easy to obtain a spuriously high reliability by choosing to analyse a sample in which, say, the behaviour never occurred or occurred on all sample points. Ideally, the test samples should be a random sample and reliability should be measured under the normal conditions of the study (Caro *et al.*, 1979).

A hypothetical example of inter-observer reliability is shown below. Two observers (A and B) each recorded the frequency of the same behaviour pattern during seven different recording sessions.

SESSION No.	1	2	3	4	5	6	7
Frequency (h^{-1})							
OBSERVER A:	23	12	34	17	24	13	37
OBSERVER B:	18	15	30	22	25	10	41

Expressed as a Pearson correlation ($n = 7$; 5 d.f.), the inter-observer reliability is: $r = +0.92$. Expressed as a Spearman rank correlation ($n = 7$), the reliability is: $r_s = +0.93$.

How reliable must a behavioural measure be before it is acceptable? No magic figure exists, above which all measures are acceptable and below which none are. Acceptability depends on several factors, including the importance of the category and the ease with which it can be measured. However, in the case of an important category that is difficult to measure, a rough guideline for acceptability might be a correlation of at least 0.7. (With a Pearson correlation of 0.7, roughly 50% of the variance in one set of scores is accounted for statistically by the other set of scores, since the coefficient of determination $r^2 = 0.7^2 = 0.49$; see 9.G.) We must stress, though, that this is only an informal guideline; some authorities would probably argue that a reliability of 0.7 is too low for any measure, no matter how important. For categories of behaviour where measurement is straightforward, reliability should be considerably better than 0.7.

Finally, note that the level of statistical significance (the 'p value') of the correlation says little about the degree of reliability, because the level of significance depends on the sample size as well as the strength of the association (or 'effect size'; see 9.D on effect size versus statistical significance). The size of the correlation coefficient, not its statistical significance, is what matters. For example, a correlation (r) of 0.5 would represent a poor degree of reliability, yet this correlation is highly statistically significant ($p < 0.01$) with a sufficiently large sample size ($n > 26$ pairs).

D Other ways of measuring reliability

1 *Index of concordance.* Another way of measuring reliability, which is particularly suited to nominal or classificatory measures (see 9.C.7), is to note whether or not there is categorical (yes or no) agreement about each occurrence of the behaviour. At the end of an observation session, two observers compare the total number of agreements (A) and disagreements (D). One measure of inter-observer reliability, known as an **index of concordance**, is the proportion of all occurrences about which the two observers agreed, i.e., A/(A + D). (The same index expressed as a percentage is sometimes referred to as the *percentage agreement*.) Other ways of calculating an index of concordance are given by Hollenbeck (1978) and Caro *et al.* (1979).

An index of concordance, rather than a correlation, need only be used if there is some reason why agreement over each occurrence of the behaviour is an important issue, or if the behaviour is measured on a nominal (or classificatory) scale (see 9.C.7). As a general rule, a measure of reliability should be calculated using the same type of measure (such as a frequency or total duration) as is used in the final presentation and analysis of the results.

2 *Kappa coefficient (κ).* The simple index of concordance described above does not take account of agreements that arise by chance alone. An index of inter-observer reliability which does allow for chance agreements is the kappa coefficient:

$$\kappa = (O\!-\!C)/(1\!-\!C)$$

where O = the observed proportion of agreements (i.e., the index of concordance, as given above); and C = the proportion of agreements that can be accounted for by chance (see Kraemer, 1979*b* ; Lehner, 1979, p. 133). For example, suppose two observers simultaneously recorded a behaviour pattern for 15 min, using instantaneous time sampling at 30-s intervals (i.e., with 30 sample points). Observer 1 recorded the behaviour on a total of 16 sample points, while observer 2 recorded the behaviour on 13 sample points. The two observers agreed (A) about the occurrence (or non-occurrence) of

the behaviour on 25 sample points and disagreed (D) or
points. The observed proportion of agreements, O = A/(
$25/30 = 0.83$. The chance proportion of agreements (C) is given by
the probability that both observers will score an occurrence (or both
score a non-occurrence) on the same sample point if their scores are
distributed randomly, i.e., $C = (16/30 \times 13/30) + (14/30 \times
17/30) = 0.23 + 0.26 = 0.49$. Therefore, the kappa coefficient, $\kappa =
(0.83–0.49)/(1–0.49) = 0.34/0.51 = 0.67$. The kappa coefficient is
considerably lower than the simple index of concordance, showing
that a substantial number of agreements may be accounted for by
chance alone. For an example of how kappa is calculated in the case
of more than one category, see Hollenbeck (1978).

3 *Kendall coefficient of concordance (W).* In studies where three or
more observers measure the same behaviour, the Kendall coefficient
of concordance (W) can be used to quantify the overall agreement
among them. W is a nonparametric statistic (see 9.F) which expresses
the degree of association among k sets of rankings (see Siegel, 1956,
ch. 9).

E Factors affecting reliability

Needless to say, many factors affect how well a category of
behaviour is measured, and these relate both to the measurement
technique and to the nature of the behaviour itself. Among the most
important factors influencing reliability are the following.

1 *Frequency of occurrence.* If a behaviour pattern occurs very rapidly
it may be difficult to record each occurrence reliably. Little can be
done about this, except to record the behaviour on film or videotape
and analyse it later in a slowed-down form. Conversely, rare
behaviour patterns may be missed altogether if observation sessions
are not long enough.

2 *Observer fatigue.* If an observation session lasts too long the
observer's ability to record accurately will be impaired through
fatigue and loss of concentration. An observer must balance the

quantity of data collected in each session against the quality of the data.

3 *Adequacy of definition.* One important factor that can often be improved is how well each category of behaviour is defined (see 3.D). If a category is not clearly and unambiguously defined then it probably cannot be recorded reliably. A common problem with protracted studies is that definitions and criteria tend to drift with the passage of time, as observers become more familiar with the behaviour and (often unconsciously) 'improve' or 'sharpen up' the definitions. This is referred to as **observer drift**, and applies to studies involving one or several observers. One way of guarding against observer drift is to measure reliability during the study and at the end, as well as at the beginning. The simplest precaution, though, is to write down the clearest possible definitions and ensure that all observers are fully familiar with these throughout the study.

F Dealing with unreliable measures

What happens if an important measure – one that is central to the question being asked – turns out to be unreliable? If statistical purity were the only arbiter then all unreliable measures would automatically be discarded. This should certainly be the fate of any unreliable measures that are also clearly irrelevant or uninformative. However, unreliable ('bad') measures are sometimes important ('right') and statistical purity must therefore be sacrificed for the sake of measuring what really matters. Two things can be done to help alleviate the problem of unreliable measures.

First, unreliable measures can sometimes be re-defined ('sharpened up'), or the measurement technique improved, to make them more reliable. For example, modifying the definition may eliminate ambiguous cases that are difficult to categorise. (Obviously, data acquired before and after a definition is changed must be analysed separately.) Similarly, if a category is difficult to record reliably using continuous recording, time sampling may give a more reliable record. For some difficult categories, reliable measurement just requires lots of practice.

Second, it is sometimes possible to combine two or more unreliable measures to produce a more reliable composite measure (see next section). This approach is entirely a matter for trial and error in individual cases, and success is by no means guaranteed.

G Composite measures

When two or more related categories of behaviour have been measured, they can sometimes be combined at the analysis stage to give a single, composite measure. This can be helpful for three different reasons.

(*a*) Combining related measures eliminates redundancy and reduces the number of categories used in the final analysis and presentation of results. Also, when the absolute frequencies of the separate scores are low on average and many individuals have scores of zero, the combined score may be more sensitive than any of its elements and easier to analyse statistically.

(*b*) Combining several unreliable measures can sometimes yield a single, reliable measure (Kraemer, 1979*b*).

In both these cases, the composite measure must have 'face validity'; that is, the separate measures must 'mean' the same and the composite measure must make intuitive and biological sense. The separate measures must therefore describe causally and/or functionally similar behaviour patterns.

(*c*) Rather than combining inter-related measures, it can sometimes be informative to combine mutually exclusive measures which are alternative expressions of a single, underlying propensity. For example, when presented with a potentially threatening stimulus, an animal may react in a number of different, mutually exclusive ways (attack, freeze, flee, etc.). If exhibiting any one of these alternative responses indicates the same underlying motivational state (fear) then pooling these measures would make sense, even though they do not overtly measure the same behaviour patterns and are not positively correlated with one another.

If the benefits of combining measures are likely to outweigh the disadvantages (such as discarding potentially helpful information), how should the component measures be chosen and, once chosen, how should they be combined? Picking the component measures is often done intuitively or on the basis of other knowledge. A more systematic approach is to inspect a measure × measure correlation matrix in order to see which measures tend to be inter-correlated (e.g., Stevenson-Hinde, 1983).

At a more advanced level, multivariate statistical techniques, such as principal component analysis, factor analysis or cluster analysis, can identify groups of measures that are statistically inter-related (e.g., Maxwell, 1977; DeGhett, 1978; Frey & Pimentel, 1978; Gottman, 1978). For example, Halliday (1976) noticed that separate measures of sexual behaviour in male newts tended to co-vary. By means of principal component analysis he was able to find a single statistical component, which he called 'libido', that accounted for the variations in several of the original measures of sexual behaviour. Note, however, that multivariate techniques make several restrictive assumptions which are commonly violated by behavioural data, and must therefore be used with considerable caution (Maxwell, 1977). For example, factor analysis should not be used if the sample size is small (less than three times the number of variables) or if variables are unreliable (Short & Horn, 1984).

The measures that are to be combined usually need to be standardised so that they have the same mean and variation. One way is to calculate for each raw value its *z* score (the score for that individual minus the mean score for the sample, divided by the standard deviation). Scores standardised in this way have a mean of zero and a standard deviation of 1.0. The composite score for an individual is then the average of the *z* scores of the separate measures. This procedure gives the same statistical weight to each measure. If different weights are to be given to the separate measures, this is best done explicitly by multiplying the *z* score of each measure by an amount that can be specified; for instance, by its loading on a principal component, obtained by principal component analysis (e.g., Stevenson-Hinde, 1983).

H **Further reading**

For an account of reliability and different ways of measuring it, see Hollenbeck (1978), Caro *et al.* (1979), Kraemer (1979*b*) or Lehner (1979, pp. 129–136). For a general account of reliability and validity, see Ghiselli *et al.* (1981, chs. 8–10). For explanations of multivariate techniques, see Maxwell (1977) and chapters in Colgan (1978).

7

Field studies

A Benefits of field studies

Some of the best and most influential studies of behaviour have been of free-living animals and people. A captive animal is usually too constrained by its artificial environment to perform even a small fraction of the activities of which it is capable. Furthermore, experimental evidence that a particular factor *can* influence behaviour may not mean that the factor *does* influence the behaviour of free-living individuals.

To observe the full richness of an individual's behavioural repertoire and understand the conditions to which each activity is adapted, the species must usually be studied in the 'field' (used in the broad sense of any environment in which individuals can range freely and interact with their own and other species). The observer notices the circumstances in which an activity is performed and those in which it never occurs, thereby obtaining clues as to what the behaviour pattern might be for (its function) and how it is controlled (its proximate causation). Useful evidence also comes from comparisons of different species of free-living animals.

A major justification for field work, therefore, is that it uncovers aspects of behaviour that would not otherwise be known about, providing the raw material from which research questions and hypotheses can be formulated. Moreover, it provides an understanding of how an animal's behaviour is adapted to the environment in which it normally lives, in the same way that its anatomical or physiological characteristics are adapted. Field studies have been particularly valuable in relating behaviour patterns to the social and ecological conditions in which they normally occur (see Krebs & Davies, 1981, ch. 2). Studies in unconstrained conditions have, therefore, been an important feature of behavioural biology.

To ask what a behaviour pattern is for is an important question in its own right, but the knowledge obtained by answering it may also be helpful for scientifically based conservation and wild-life management. Moreover, the functional approach can greatly aid those who are primarily interested in the study of behavioural mechanism, leading the investigator to the major variables controlling a behaviour pattern. Being able to distinguish the important causal factors is extremely useful when designing experiments – in which, inevitably, only a small number of independent variables are actually manipulated while the others are held constant or randomised (see 2.C).

Even when the focus is on how behaviour is controlled, field studies may be essential because activities of interest, such as immigration or ranging, simply are not exhibited in captive conditions. Similarly, animals may only show certain forms of behaviour if they live in normal social groups and have developed in a natural environment. For example, Seyfarth & Cheney's work, which showed that vervet monkeys in the wild give different types of alarm call in response to different types of predator (leopards, eagles and snakes), would not have been practicable in captive conditions (see Seyfarth *et al.*, 1980).

Good studies of animals living in natural conditions do not necessitate going to exotic places. Some of the best field work has been carried out on temperate species in the gardens, ponds and woodlands near major centres of human population. Nonetheless, the biological richness of the tropics or the opportunity to study animals in relatively undisturbed habitats provide powerful motives to venture further into remote areas.

B Difficulties of field studies

Many field projects fail to anticipate major practical problems. An animal under observation may frequently disappear from view, wrecking the best-laid plans for systematic recording over a fixed period. Similarly, distributing the choices of focal individual randomly or evenly among members of a population may not be easy because some identified individuals are difficult to find (see 4.A on sampling rules). Bad weather may make observation impossible. The animals may prove to be much more shy than had been expected and require months (and in some cases years) of habituation before they will allow a human observer near enough for useful observations to be made. In general, the conditions for recording behaviour in the field are rarely ideal and high-quality data are not easy to collect (see, for example, Aldrich-Blake, 1970).

The days are over when a field worker could confidently suppose that good descriptions of a species obtained from one habitat could be generalised to the same species in another set of environmental conditions. Consequently, the field worker may find that a great deal of effort has been put into describing a special case, and the study may only start to make sense when the same species has been studied in a number of different habitats (e.g., Berger, 1979).

Studying animals in the field can sometimes involve both physical and mental hardship because of the need to work in remote places with harsh climates. The field worker may have to live for extensive periods in difficult circumstances, facing isolation, possible ill health, poor diet and occasional physical danger. The advice and facilities which are taken for granted in an academic environment are rarely available. Problems with logistics and bureaucracy may mean that less research is done than expected, because everything takes more time. On top of all this, field workers returning home from long spells of work in remote places may face considerable disorientation and personal difficulty (see the report on the problems of field workers by Hinde, 1979*a*).

C Some practical advice

The special problems of studying behaviour in a natural environment mean that the field worker often has to make compromises between what is ideal and what is practicable. It is not easy to produce general rules about such compromises, since they will depend so much on the species being studied and the conditions in which it lives. Choice of the right species is especially important in field work (see 1.F) and we assume that considerable thought will already have been put into this matter before embarking on a study. What follows are some suggestions about how to get the most out of the field work.

(*a*) The importance of becoming thoroughly familiar with the species and the study site cannot be over-emphasised. Most field workers find that their final results contain little from their first few months spent in the field. The preliminary period of exploration may well involve replacing seemingly excellent theoretical notions of what *should* be done by a more down to earth appreciation of what *can* be done. All research involves an element of opportunism, grabbing chances when they are presented. But field work, in particular, requires flexibility and a readiness to change plans when a course of action is frustrated by unexpected difficulties. Deciding the appropriate moment at which to switch from the initial exploratory phase to more tightly focused research can be difficult and is best done after discussion with other experienced field workers.

(*b*) People intending to work in a remote area for long periods of time are well advised to take regular breaks during the course of their study. The advantages of a spell away from the study site, opportunities to discuss data with colleagues and, generally, to see their work in a broader perspective almost invariably outweigh the loss of time spent in collecting data.

(*c*) In a desire to gather usable results under the sometimes difficult conditions of a field study, it can be especially tempting to pick reliable measures, even if this means measuring something unimportant – in other words, to choose *good* measures rather than *right* measures (see 6.A). Field work provides unique opportunities for observing important aspects of a species' behaviour and biology, and observation methods need to be sufficiently flexible so that rare events can be properly recorded (see 4.A).

(*d*) Scan samples can be unreliable, particularly when the visibility of individuals is variable (see 4.A). Watching a single individual continuously for a specified period (focal sampling) overcomes most of these difficulties, even though smaller sample sizes are obtained. Focal sampling also provides greater insight into the problems faced by the individual and the ways it solves them.

(*e*) Focal sampling in the field usually requires that the observer is able to recognise individuals reliably (see 8.B). Difficulties can arise if, as often happens, a focal individual moves out of sight. If the subject cannot be followed, it is a good idea to have a firm rule for giving up an unsuccessful search after a specified period of time. Data collected on an individual for a shorter period of time than was allotted to it can certainly be used, but must be corrected for the time that it was in view. In general, keep track of how much each identified individual in a study area has been observed. It will then be easier to ensure that observations are distributed relatively evenly among the (visible) members of the population.

(*f*) Animals that are initially wary generally become used to an observer, although rates of habituation depend greatly on how much the animals are disturbed by other humans (see 2.B). Habituation is often a desirable goal because it makes observation so much easier, but remember that an animal that is less likely to run away from a scientist may also be less likely to run away from a poacher (see 10.E on ethics). In order to allow the animals to become familiar with an observer, they must not be alarmed. Approach should be indirect and accompanied by innocuous actions. The observer should avoid sudden movements, staring at the animals or following them immediately if they move away. Although habituating the subjects may take a long time, it should not be hurried; a small amount of time spent on habituation each day is much more effective than a longer period at less frequent intervals.

(*g*) Do not forget that, even with seemingly well-habituated animals, the observer may unwittingly influence the subject's behaviour (see 2.B) and these effects may not be immediately obvious. The observer's presence may introduce bias; for example, by suppressing some activities (such as play) but not others, or by scaring away young animals but not old ones (or vice versa). More subtly, habituated subjects may ignore the observer while their predators (or prey) keep their distance.

(*h*) Estimating distances is a skill that should be practised and repeatedly checked against objective measurements. For instance, it can be instructive to estimate the distance between two objects and then measure the actual distance between them. Inexperienced observers can be alarmingly inaccurate in making such estimates, but will never realise this unless they occasionally check their estimates against actual measurements. It is best to practise on objects that are as far away as are the animals when normally studied. Sometimes it is possible to assist estimations of distance by placing pegs at known distances apart in the areas where the animals are likely to be seen. Another useful aid to estimating distances is the length of the animals themselves. Indeed, many field workers use animal body-lengths as a standard unit of measurement.

(*i*) Estimating the numbers of individuals in large groups can be practised using photographs or drawings with animal-shaped blobs on them. A rapid guess can then be checked against the actual number on the sheet. However, practice is best obtained in the field. A quick estimate of the number of individuals in a real group of animals can be made and the actual number then counted, either by counting them on the spot or, less satisfactorily, by taking a photograph and counting them later.

D Field experiments

The point of doing an experiment is to distinguish between alternative explanations or hypotheses. Field observation of natural variation can also achieve this goal if, for example, naturally occurring events demonstrate associations between variables that previously seemed unrelated, or break associations between variables that previously seemed bound together. Nonetheless, an observer may have to wait an inordinate amount of time for the lucky event, and good experiments can sometimes be performed in the field.

For example, tape-recordings of predators or conspecifics (such as offspring or potential mates) can be played to free-living animals and their responses measured (e.g., Cheney & Seyfarth, 1982); dummies of different designs can be used to gauge responsiveness to a particular shape or colour; food can be provided in order to influence the time spent on non-feeding activities such as parental behaviour, and so forth. The possibilities for such work are considerable. However, it has to be remembered that badly designed experiments are a waste of time. Moreover, all experiments are liable to be disruptive to the animals (see 10.E on ethics) and interfere with those qualities that field studies are best equipped to investigate, namely the natural character of behaviour.

Field experiments can be used to understand how an animal's behaviour is controlled, but they can also test functional hypotheses. For example, Tinbergen and colleagues (1962) wanted to know why the ground-nesting black-headed gull removes the pieces of egg shell from its nest site after the chick has hatched. A number of different hypotheses initially seemed plausible: the chick might injure itself on the sharp edges of the shell; the shell might be a source of disease by harbouring micro-organisms; an egg might get trapped under a piece of shell, forming a double shell from which the hatching chick could not escape; the pale inner surface of the shell might attract predators visually; or the smell might attract predators by olfactory cues. Some of these hypotheses could be excluded using comparative evidence. Another gull, the kittiwake, nests on cliffs where it is not vulnerable to predators and does not remove the egg shell from its nest. This suggested that the first three possibilities were unlikely to be of major importance. A simple field experiment, which involved placing shells at different distances from the nest, was used to show that broken shells do indeed attract predators to the black-headed gull's nest. The egg is cryptically coloured on the outside, but the inside is pale and therefore easy for an airborne predator such as a crow to spot. The field experiment confirmed that nests with broken shells lying near them were more likely to be raided (see also Drickamer & Vessey, 1982, pp. 27–30).

E **Further reading**

The role of field work in the study of animal behaviour is discussed by Baerends (1981) and Drickamer & Vessey (1982, ch. 3). Good examples of field study methods are found in Altmann (1980), Clutton-Brock *et al.* (1982), Seyfarth *et al.* (1980) and Cheney & Seyfarth (1982). Some problems of bias in field observations are outlined by Aldrich-Blake (1970). The many problems, both physical and mental, that are encountered by some field workers are reviewed by Hinde (1979*a*). See also Schneirla (1950), Hailman (1973), Lott (1975) and Lehner (1979). Although this chapter has been concerned with field studies of animals, many of the same issues arise in observational work on the behaviour of humans (see, for example, Hutt & Hutt, 1970; Borgerhoff Mulder & Caro, 1985).

8

Social behaviour

A Defining a group

Many of the most interesting and important aspects of behaviour involve social interactions between individuals. In this chapter we give some examples of how such patterns of social behaviour may be described. We start by considering what constitutes a group: where does one stop and another start?

In practice, the rules for defining groups are usually implicit: groups can often be defined intuitively by assessing how the animals are distributed in space and observing the relative distances between individuals. Those below a certain distance are regarded as being within a group, and those above it as outside. This informal approach can cause difficulties if the groups are not tightly clustered or if individuals spend only some of their time together. It is important, therefore, to make the rules for defining a group as explicit as possible, and to use carefully chosen criteria when deciding on the critical distances and minimum time spent with others that define membership of a group. A distinction is sometimes drawn between **groups** – associations whose composition is known – and **parties**, which are aggregations whose membership is uncertain (Crook, 1970).

One way of deciding on the criterion distance for defining membership of a group or party is illustrated by the work of Clutton-Brock *et al.* (1982, appendix 12) on red deer. They started with the southernmost deer and estimated by eye the distance of its nearest neighbour in a northerly direction. (The nearest neighbour was defined as the one whose head was closest to the head of the target animal.) They did the same for the next most southerly animal and worked systematically through all the deer that were visible. They discovered that the nearest-neighbour distances were bimodally distributed, with most neighbours being found either within 40 m or more than 60 m away, and few neighbours being found between 40 and 60 m. The cut-off point therefore appeared to be a distance of about 50 m, so deer were treated as being in a party if the distance between them was less than 50 m. The usefulness of this criterion was confirmed by showing that, on average, 90% of the deer in the same party were engaged in the same activity, whereas only 56% of deer in different parties did the same thing at the same time.

We should emphasise that this is but one of many ways to define a group or party, and that the problems of definition vary hugely between species. What is appropriate for red deer may not, for example, be appropriate for fish or children.

B **Identifying individuals**

For many studies, being able to identify individual subjects is essential. Focal sampling (4.A), for example, is virtually impossible without a reliable identification method. Furthermore, the theoretical interest of a study is usually enhanced when differences in the behaviour of known individuals are recorded. Only by identifying and watching individuals has it become clear that all individuals in a species do not behave in the same 'species-typical' way. On the contrary, marked intraspecific variations in behaviour have been described for many species, and many of these differences make functional sense (see Slater, 1981; Krebs & Davies, 1981, ch. 8; Davies, 1982; Trivers, 1985, ch. 9).

In the laboratory, identification of individuals by rings, tags, tattoo marks, painting the skin or fur, toe-clipping, ear-punching, collars, belts, freeze-branding, fur-clipping and so forth does not usually offer major practical difficulties (Lane-Petter, 1978; Twigg, 1978). However, it is important to realise that marking an individual may alter its behaviour or that of other individuals. To give one example, an experiment showed that coloured plastic leg bands worn by zebra finches affect how attractive they are to members of the opposite sex. Female zebra finches prefer males wearing red leg bands over unbanded males, while males prefer females with black leg bands. Both males and females avoid members of the opposite sex wearing green or blue leg bands (Burley *et al*., 1982; see also Trivers, 1985, p. 256). These results clearly show that for zebra finches, and probably many other species, identification markings can have an important effect on behaviour.

In the field, trapping and marking can present formidable problems. Capturing animals using traps, nets or stupefying drugs can be difficult (Eltringham, 1978), and some forms of marking do not last long under field conditions. Careful thought should be given to minimising the distress caused by marking animals, for scientific as well as ethical reasons (see 10.E).

In addition to visual markings, modern techniques enable animals to be tracked over large distances using miniature radio transmitters or sources of radioactivity attached to the subject (Linn, 1978; Amlaner & Macdonald, 1980; Mech, 1983).

In some species, individuals have naturally distinctive markings. For example, zebras' stripes are like human finger prints, no two individuals being identical. Similarly, gorillas' noses, elephants' ears, the whisker spots of lions, cheetahs' tails and the bills of Bewick's swans (to list only a few) are highly variable, enabling experienced observers to recognise individuals (e.g., Scott, 1978). Many animals living in the wild acquire distinctive marks through injury, such as torn ears, damaged tails, scars or stiff limbs. Again, features such as these can be used to distinguish one individual from another. Identifying animals by naturally occurring features can be difficult and requires patience and practice (Pennycuick, 1978), but it is the best approach in terms of minimising suffering and disruption.

When, as is often the case, experienced observers are convinced that they can recognise individuals without recourse to written records, some public demonstration of their ability is advisable. One technique is to photograph the animals as they are simultaneously identified by the observer. The tester then removes from each photograph additional environmental cues which might be helpful in identification, and records the identity of each individual. Days or weeks later, the tester presents the observer with the pictures in random order, and asks the observer to name each individual. In one such test, an observer who could individually recognise some 450 adult Bewick's swans identified the individual swan correctly in 29 out of 30 photographs (Bateson, 1977).

C Dominance hierarchies

In many social species, the relationships between pairs of individuals are asymmetrical. One individual will consistently supplant the other when they compete for a valued resource such as food, shelter or a mate, or may simply cause it to move away when they meet. If the numbers of such occurrences are recorded for every pair in the group, it often becomes apparent that one individual tends to supplant all other animals, whereas another is supplanted by all others. In between the top- and bottom-ranking animals are animals that supplant some but are supplanted by others. The overall arrangement of dominant and subordinate individuals in the group is referred to as a **dominance hierarchy**.

To derive a dominance hierarchy from observations of the interactions between individuals, the numbers of supplants between pairs are arranged in a matrix as shown in the upper matrix in Fig. 8.1. The order is then arranged so that the individual that is never supplanted is at the top and the one that is always supplanted is at the bottom. The other animals are re-arranged in order until the minimum number of supplants appears on the left-hand side of the diagonal. The final order is shown in the lower matrix in Fig. 8.1, where the individuals have been arranged into a dominance hierarchy.

If, as in this example, all individuals in the group can be arranged in strict order of dominance (C dominates A, A dominates D, D dominates E and E dominates B), then the dominance hierarchy is said to be **linear**. In reality, however, few dominance hierarchies are perfectly linear. Sometimes **dominance reversals** may occur, when a subordinate wins an encounter with a normally dominant individual. Furthermore, for a hierarchy to be perfectly linear all dyadic relationships must be **asymmetric**, whereas in some groups two or more individuals may have equal status. Finally, in a perfectly linear hierarchy all possible triadic relationships must be **transitive**. (In a transitive relationship, if A dominates B and B dominates C then A must also dominate C.)

Fig. 8.1. Constructing a dominance matrix. Hypothetical example of how five individuals (A,B,C,D,E) are rank-ordered on the basis of dyadic interactions, according to who supplants whom. The upper matrix shows the number of occasions on which one individual supplanted the other, for all possible combinations. In the lower matrix the individuals are re-arranged so that the dominant individual (C, who supplants all the others) is at the top. The dominance hierarchy in this example (C > A > D > E > B) is perfectly linear, with no reversals. Hence, Landau's index ($h = 0.1 \times [4 + 1 + 0 + 1 + 4]$) for this matrix is 1.0.

Number of occasions when individual is supplanted

	A	B	C	D	E
A	—	21	0	29	11
B	0	—	0	0	0
C	22	11	—	8	18
D	0	11	0	—	6
E	0	2	0	0	—

	C	A	D	E	B
C	—	22	8	18	11
A	0	—	29	11	21
D	0	0	—	6	11
E	0	0	0	—	2
B	0	0	0	0	—

Number of occasions when individual supplants other

Landau's index of linearity (h) provides a measure of the degree to which a dominance hierarchy is linear (see Chase, 1974; Bekoff, 1977; or Lehner, 1979, p. 217). The index is calculated as follows:

$$h = \frac{12}{n^3 - n} \cdot \sum_{a=1}^{n} \left(v_a - \tfrac{1}{2}(n-1) \right)^2$$

where n is the number of animals in the group and v_a is the number of individuals whom individual a has dominated. The index ranges from zero to 1.0, with a value of 1.0 indicating perfect linearity. Values of h greater than 0.9 are generally taken to denote a strongly linear hierarchy. The hierarchy shown in Fig. 8.1 contains no dominance reversals, symmetrical dyadic relations or intransitive triadic relations and, accordingly, has a Landau's index of 1.0.

The procedure of arranging animals to form a dominance hierarchy, as shown above, can be deceptive, because there is a surprisingly high probability that a set of data can be arranged to form a linear dominance hierarchy when none exists in reality (Appleby, 1983). This is especially likely when the group is small and the observer, having no independent knowledge of the animals' relationships, *assumes* the existence of a linear hierarchy and juggles the data until the best hierarchy is obtained. Appleby (1983) has described a method for calculating the probability that an apparent dominance hierarchy will arise from a set of observations when none exists in reality. A perfectly linear hierarchy comprising five or less individuals can be obtained by chance (that is, when the underlying dominance relationships are actually random) with a probability of more than 0.05. Thus, only data from groups of six or more individuals can be shown to have a statistically significant linearity. If a few imperfections are permitted in the hierarchy, then the probability of a roughly linear hierarchy arising by chance are even greater. For example, in a group of six individuals, an apparent hierarchy with one dominance reversal has a 0.18 probability of occurring in a randomly ordered set of data (Appleby, 1983). Of course, if repeated observations of social interactions consistently indicate the same hierarchy, then it can be accepted with greater confidence.

In its simplest form, as described above, the index of dominance status assigned to each individual is its rank in the hierarchy. Thus, dominance is measured on an ordinal (ranking) scale, which means that the magnitude of the difference in dominance status between two individuals cannot be quantified (see 9.C.7). However, Boyd & Silk (1983) have developed a way of measuring dominance on an interval scale, using a method of paired comparisons. With this index of dominance, the difference in dominance between two individuals can be quantified and tested for statistical significance. Furthermore, since measurement is on an interval scale, parametric statistics can be used (see 9.F). This method is particularly useful for describing dominance hierarchies that are not highly linear, and can also be used for analysing any asymmetric interactions that involve an actor and a recipient, such as grooming or food-sharing.

Finally, we urge caution in interpreting dominance hierarchies. One common error is to over-generalise the meaning of a dominance hierarchy, by treating the dominance status of each individual as though it were a fixed and general characteristic of that individual. On the contrary, dominance relations are often fluid and capable of rapid change. Dominance relations sometimes have a geographical element, an individual's rank increasing towards the centre of its home range. In such cases, dominance reversals must be interpreted in the light of where the encounter occurs with respect to the two individuals' home ranges. Furthermore, a dominance hierarchy derived from one measure (such as competitive interactions over food) is not always the same as the hierarchy derived from a different measure (such as competition for mates).

D Indices of association

It may be useful to measure the extent to which two individuals, A and B, associate with each other. Any measure of association between the two must not only take account of the number of separate occasions that A and B are seen together, but also the number of separate occasions that A is seen on its own and the number of separate occasions that B is seen on its own. In order to obtain such a measure, it is necessary to define what is meant by 'separate occasions' and 'together'. For example, it is no use treating scan samples of slow-moving animals taken at 30-s intervals as being separate, since one observation is effectively the same as the next (see 2.E on independence of measurements). The definition of 'together' will depend on the study: it might require that the animals are in bodily contact; that they are within one body-length of each other; or even that they are merely within visual contact.

Several **coefficients of association** have been devised (e.g., Clutton-Brock *et al.*, 1982, p. 47). One simple and straightforward index is as follows:

$$\text{Index of association} = N_{AB} / (N_A + N_B + N_{AB})$$

where N_{AB} = number of occasions A and B are seen together; N_A = number of occasions A is seen without B; and N_B = number of occasions B is seen without A. This index has the merit that all scores are distributed between 0 (no association) and 1.0 (complete association). A score of 0.5 means that the two animals are seen apart as often as they are seen together.

For most indices of association, the chance level of association is not easily calculated and can only be estimated roughly from the likely constraints on the random movements of two independent bodies. Clearly, the larger the area in which individuals can move, the less likely they are to meet by chance. In the absence of a chance level of association, the use of an association index is restricted to making comparisons (between, say, the two sexes or between different age classes).

Associations between large numbers of individuals within a group can be analysed using a multivariate statistical technique known as cluster analysis (Morgan *et al.*, 1976; Maxwell, 1977).

E   Maintenance of proximity

Two individuals, such as a mother and her offspring, may spend a great deal of time together. An important measure of their relationship is the extent to which their proximity is due to the movements of one member of the dyad rather than the other. This is measured by counting the occasions on which one member of the dyad approaches or leaves the other while the other member remains still. (Strictly speaking, both individuals could be moving. The precise question is: which individual's movements produced a crossing of an imaginary circle of specified radius around the other individual?) The number of occasions when the pair came together or were separated as the result each individual's movements can then be obtained.

A measure of the extent to which individual A has been responsible for maintaining proximity between itself and individual B can then be calculated as follows:

A's responsibility for proximity
$$= U_A/(U_A + U_B) - S_A/(S_A + S_B)$$

where U_A = number of occasions when pair were united by A's movements; U_B = number of occasions when pair were united by B's movements; S_A = number of occasions when pair were separated by A's movements; and S_B = number of occasions when pair were separated by B's movements. The index ranges from -1.0 (B totally responsible for maintaining proximity) to $+1.0$ (A totally responsible). A value of 0 indicates that A and B were equally responsible for maintaining proximity. As with measures of association (previous section), the major value of such an index lies in making comparisons. For instance, the index has been used to show how the role of a rhesus monkey mother in maintaining contact with her offspring steadily declines as the offspring gets older (Hinde & Atkinson, 1970).

F Further reading

Hinde (1979*b*, 1983) and Cairns (1979) cover many aspects of the study of social behaviour in humans and other primates. See also Broom (1981, ch. 10). For details about the capture and marking of animals, see chapters in Stonehouse (1978) or Amlaner & Macdonald (1980). Bekoff (1977), Lehner (1979, ch. 9), Appleby

(1983) and Boyd & Silk (1983) deal with methodological issues concerning dominance hierarchies. For a discussion of dominance, see Richards (1974), and Bernstein (1981) and accompanying commentaries.

9

Statistics and data analysis

A **General advice**

Our general advice is not to become obsessed by statistical techniques, nor too cavalier in their use. Statistical analysis is a tool to help answer questions, and should be the servant rather than the master of science. The physicist Lord Rutherford was over-stating this point when he wrote: 'If your experiment needs statistics, you ought to have done a better experiment'; biological systems can be extremely complex and statistical analysis is often essential for understanding what is going on. Nonetheless, excessively complicated statistics are sometimes used as a poor substitute for clarity of thought or good research design.

B **Exploratory versus confirmatory analysis**

Statistical techniques are used for two quite different purposes: **exploratory** data analysis, and hypothesis-testing or **confirmatory** analysis. Most conventional statistics textbooks deal mainly with confirmatory analysis.

1 *Exploratory data analysis* (or descriptive statistics) includes the essential (but often neglected) processes of collating, summarising and presenting results, and searching through them so as to obtain the maximum amount of information. This is especially important when the results are complex or the hypotheses vague. Exploratory analysis provides a way of learning from results and generating new hypotheses from them. On no account, however, should hypotheses

be generated and then tested using the *same* data, 'a procedure tantamount to offering to bet on a horse race after the finish of the race' (Kraemer, 1981).

One of the simplest and most fruitful types of exploratory analysis is plotting results in the form of a suitable graph, histogram or scatter plot. It is always wise to plot results and inspect them visually before plunging into confirmatory statistical analysis. As a general rule, graphical summaries are much more informative than tables of figures (see 10.C on plotting data). It is also helpful to summarise data in the form of summary descriptive statistics, such as means or medians and standard deviations or ranges, before hypothesis-testing is carried out. (Remember that if a variable is normally distributed, approximately two-thirds of all scores (68%) will fall within ± 1 standard deviation of the mean, roughly 95% within ± 2 standard deviations, and virtually all scores (99.7%) within ± 3 standard deviations of the mean.)

Exploratory data analysis (EDA) techniques provide many useful ways of investigating data rapidly without recourse to a computer (Tukey, 1977; Velleman & Hoaglin, 1981). If, on the other hand, large amounts of data are to be analysed using a computer, exploratory analysis is still a fruitful first step. Some of the statistics software packages available for mini- and microcomputers, such as MINITAB, include facilities for EDA techniques (Ryan *et al.*, 1985).

2 *Confirmatory data analysis* (hypothesis-testing, or inferential statistics) covers the conventional 'testing' of empirical data; that is, calculating the probability that the observed result is consistent with a **null hypothesis** (see 9.C) such as 'there is no difference between the mean scores of the two groups', or 'there is no association between the two sets of scores'. If this probability is lower than a predetermined level (usually 0.05), then the null hypothesis is rejected. (Incidentally, 'confirmatory' analysis is something of a misnomer since, according to many philosophers of science, the aim of experiments should be to refute rather than confirm hypotheses.)

The main purpose of hypothesis-testing is to provide a publicly understood way of specifying how much confidence can be placed in an apparent effect, such as a difference or correlation. Confirmatory data analysis is central to studies of natural variation as well as experiments, since it includes tests of hypotheses about correlation (see 9.G).

Confirmatory data analysis can be performed on a mini- or microcomputer using one of several standard statistics software packages (such as SPSS, MINITAB or SAS). However, we would urge caution in the use of such computer packages. Just because it is easy to do so using a computer, data should not simply be fed in and analysed in every possible way without prior thought about the questions being asked. Moreover, the investigator should become familiar with the data through exploratory analysis before they are fed into a computerised 'black box' for confirmatory analysis.

Data analysis is not a purely mechanical exercise. A set of data can usually be analysed in many different ways and, as findings are compared with hypotheses and expectations, new ideas arise which lead to the analysis of the data in other ways. If the data are rich and complex, the process of analysis and interpretation may be slow. Nonetheless, it is a very important phase in many research projects.

C Some statistical terms

1 *The null hypothesis* (Ho) is the baseline assumption made when testing a hypothesis, against which the outcome is compared. The null hypothesis is usually that there is no effect; for example, that there is no significant difference or correlation between two sets of data.

2 *The level of statistical significance* (α) is the probability of obtaining the observed result, or one more extreme, if the null hypothesis were true. In other words, α is the probability that the observed effect (such as a difference or correlation) arose by chance alone (through sampling error) and that there is no real underlying effect. If this probability falls below a predetermined critical level of significance, which is usually set at 0.05, then the null hypothesis is rejected. Thus, with the critical level of significance set at 0.05, the probability of rejecting the null hypothesis when it is in fact true (i.e., scoring a 'false positive' by concluding falsely that there is an effect) is less than one in twenty. It may be appropriate to set the level of significance at 0.01 or 0.10 rather than 0.05, depending on how important it is not to make the occasional false inference.

3 *Type I and Type II errors.* Rejecting the null hypothesis when it is in fact true (scoring a 'false positive' by concluding there is an effect when none exists) is referred to as a **Type I error**. Accepting the null hypothesis when it is false (scoring a 'false negative' by failing to detect a real effect) is called a **Type II error.**

4 *The power* of a statistical test is the probability of rejecting the null hypothesis when it is false; in other words, the probability of finding a real effect. Power is given by $(1-\beta)$, where β is the probability of a Type II error (see above). The greater the power of a test, the more likely it is that a real effect, such as a difference or correlation, will be detected. Statistical power can be increased either by increasing the sample size (n) or by improving the research design; for example, by decreasing measurement error.

5 *One-tailed and two-tailed tests.* If a prior prediction is made about the *direction* of an effect – for example, 'the mean score for the experimental animals is greater than for the control group' – then the test is **one-tailed**. Alternatively, if no direction is specified in advance ('the scores are different'), then the test is **two-tailed**. For instance, suppose we wished to test whether there is a significant difference in weight between males and females. If the initial hypothesis is simply that there is a difference, then the test is two-tailed. If, however, existing knowledge or theory firmly predicts the direction of the difference (for example, females heavier than males), then the test is one-tailed.

We must stress that for a one-tailed test to be used, the prediction about the direction of an effect must be a real prediction, made *before* the results are obtained. In most cases, a two-tailed test should be used.

One-tailed and two-tailed tests must be distinguished when specifying a level of significance or when looking up the critical values for a test statistic. The critical value of a test statistic (such as Student's *t* or a correlation coefficient) at the 0.05 level for a one-tailed test corresponds to the 0.10 level for a two-tailed test. Thus, an effect which just fails to be significant at the 0.05 level (i.e., $0.05 < p < 0.10$) for a two-tailed test can be made to appear significant simply by changing the 'hypothesis' so that the test becomes one-tailed. Changing from a two-tailed to a one-tailed test after the result has been obtained is downright dishonest. One-tailed tests should only be used if there are genuine *a priori* reasons for predicting the direction of a difference or correlation.

6 *Parameters and variables.* In general usage, a **parameter** is any quantity which is *constant* in the case being considered, but which may vary between different cases. 'Parameter' is sometimes used incorrectly as a synonym for *variable*, when referring to a behavioural category or some other measure. In statistical terminology, a parameter is a numerical characteristic (such as a mean or variance) describing the entire population. Parameters are usually estimated from samples rather than measured directly. Parametric statistical tests (see 9.F) are so called because they make various assumptions about population parameters, such as means, variances and frequency distributions.

7 *Levels of measurement.* Measurement means assigning numbers to observations according to specified rules. Four different **levels of measurement** are distinguished, in ascending order of strength of measurement:

(*a*) If observations are simply assigned to mutually exclusive, qualitative classes or categories (such as male/female; primiparous/multiparous; active sleep/quiet sleep/awake; or type A/type B/type C/ type D, etc.), the variable is measured on a **nominal** (or classificatory) scale.

(*b*) If the observations can also be ranked along a scale according to some common property (for example, A > B > C >...), then the variable is measured on an **ordinal** (or ranking) scale.

(*c*) If, in addition, scores can be placed on a scale such that the distance between two points on the scale is meaningful – i.e., the *difference* between two scores can be quantified – the variable is measured on an **interval** scale. The zero point and unit of measurement are arbitrary for an interval scale. A temperature measured in degrees Celsius or degrees Fahrenheit is measured on an interval scale.

(*d*) The highest level of measurement is attained when the scale has all the properties of an interval scale but also has a true zero point. This is referred to as a **ratio** scale since, unlike on an interval scale, the ratio of any two measurements is independent of the unit of measurement. Mass, length and time are measured on ratio scales. For example, the ratio of two periods of time is the same whether they are measured in seconds or days. True frequencies, durations and latencies are measured on ratio scales.

Assigning numbers to a category of behaviour does not necessarily mean that the behaviour is measured on an interval or ratio scale. For example, rating an individual's aggressiveness on a scale of 0 to 5 would not constitute measurement on a true interval scale, since there is no reason to assume that the difference between scores of, say, 1 and 2 is the same as the difference between scores of 4 and 5. Scores can be ranked (see below), but the differences between scores may not be meaningful. Thus, measurement would be on an ordinal scale.

8 *Ranks.* When a number of measurements are arranged in order according to some common quality (i.e., on an ordinal scale), they are said to be ranked. The number assigned to each measurement is its **rank**, while the arrangement as a whole is called a **ranking**.

9 *Population and sample.* The **population** is the entire set of items or individuals (animals, events, humans, adult male rhesus monkeys, middle-class female university students, etc.) under consideration and about which statistical inferences are to be made. The population is distinguished from a **sample**, which is the particular group or sub-set of entities selected from the population for measurement. Measurements are made on a sample (for example, 10 rhesus monkeys from a laboratory colony of 30) and these measurements are then used to draw statistical inferences about the population as a whole (for example, the whole colony or all rhesus monkeys living under comparable conditions). It can often be difficult deciding how far a given set of results can be generalised – in other words, specifying the population to which the sample results refer. The temptation is usually to over-generalise results; for example, by implying that the results of a laboratory experiment on learning in Norway rats can be applied directly to the process of learning in all animals.

A **random sample** is one where each and every member of the population had an equal probability of being selected for measurement. Truly random sampling is often referred to, but seldom achieved in practice. A **representative sample** is one which has the same broad characteristics as the population (for example, the same distribution of ages and weights, or the same sex ratio). A sufficiently large random sample will, on average, also be a representative sample. A **haphazard sample** is one where the sample is chosen according to an arbitrary criterion such as availability or visibility. In many studies, supposedly 'random' samples are often more like haphazard samples, and may or may not be representative of the population. Other sampling schemes are also possible – for example, stratified random sampling (see Cochran, 1977; Snedecor & Cochran, 1980, ch. 21).

D Effect size versus statistical significance

A crucial distinction exists between the magnitude of an effect – such as the size of the difference between the scores of two samples, or the size of the correlation between them – and its statistical significance. The magnitude of an effect, such as a difference or correlation, is referred to as the **effect size** (or, in medical parlance, as the 'clinical significance'). Effect size and statistical significance are quite separate issues and the level of statistical significance does not, as is often supposed, directly measure the magnitude or scientific importance ('clinical significance') of the observed result. An observed effect, such as a correlation or a difference, can be very small yet highly statistically significant, provided the sample size is large enough.

Effect size and statistical significance are frequently confused when interpreting correlation coefficients, especially when a correlation is small but statistically significant (see 9.G). Even though the strength of an association may be paltry, with a sufficiently large sample size the correlation can nonetheless be highly statistically significant. Similarly, the mean scores of two samples may differ by only a tiny amount yet, with a large sample size, this difference can still be highly statistically significant.

Whilst it is obviously essential to distinguish clearly between statistically significant and non-significant results, the level of significance by itself provides little useful information. For this reason we strongly recommend that results should include information on the effect sizes (for example, mean scores and standard deviations, or correlation coefficients), not just their statistical significance. For example, when quoting a significant difference between two groups, give the actual mean scores (plus associated measures of variation) and sample sizes. Simply stating that 'the mean score for the experimental group was greater than that for the controls ($p < 0.05$)' is insufficient, as it gives no idea of how large this difference was. Similarly, when quoting a significant correlation, state the actual correlation coefficient and sample size (or degrees of freedom), not merely the level of significance..

If a study produces a result that borders on statistical significance (e.g., $0.05 < p < 0.10$), then it may be wise to increase the sample size in order to reach a definite conclusion about whether the effect is a real one or not. This should be done by deciding in advance by how much the sample size should be increased. The practice of making piecemeal increases in the sample size until the results 'reach significance' is dangerous, since it capitalises on the chances of obtaining a spurious effect and committing a Type I error (9.C.3).

E **The procedure used in hypothesis-testing**

In this section we shall briefly outline the steps involved in statistically testing the simplest type of hypothesis; for example, that there is a difference or correlation between two measures. Broadly speaking, the aim of hypothesis-testing is to decide whether an observed effect reflects a real effect or arose through chance factors such as sampling error (see 9.I).

(*a*) *Specify a null hypothesis* (for example, 'no difference between the experimental and control groups' or 'no correlation between measures *A* and *B*') and an alternative hypothesis (for example, 'the mean for the experimental group is greater than that for the controls' or 'variable *A* is positively correlated with variable *B*'). State whether the test is one-tailed or two-tailed (see 9.C.5).

(*b*) *Select the appropriate test statistic*, which means choose the right statistical test. For example, Student's *t* test, the Mann-Whitney *U* test and the Wilcoxon matched-pairs test are used for testing the significance of a difference between two samples, although under different circumstances (see 9.K). Spearman and Pearson correlations are also test statistics, since it is possible to test hypotheses about correlations as well as differences.

(*c*) *Set the criterion level of statistical significance* (α), below which the null hypothesis ('no effect') will be rejected. This is usually set at 0.05.

(*d*) From the empirical results, *calculate the value of the test statistic*; for example, the value of Student's *t*, Mann-Whitney *U*, or a correlation coefficient.

(*e*) *Look up the critical value of the test statistic* for the particular sample size and level of significance used. (Check whether the table shows values for one-tailed or two-tailed tests and adjust the level of significance if necessary.) If the value of the test statistic falls within the region of rejection (e.g., $p < 0.05$) then reject the null hypothesis. If the test statistic falls outside the region of rejection then the null hypothesis is *provisionally* accepted; it has not, however, been proved to be true.

F   Parametric versus nonparametric statistics

Statistical tests are of two basic types: parametric or nonparametric.

1 *Parametric tests,* such as Student's *t* test, analysis of variance, linear regression and Pearson correlation, are usually the most powerful tests (as defined in 9.C.4). However, as their name suggests, parametric tests are based on certain assumptions about the nature of the population from which the sample data are drawn. Parametric tests generally require the following assumptions to be made.

(*a*) The data follow a normal distribution (the assumption of *normality*).

(*b*) Samples or sub-groups have approximately equal variances (*homogeneity of variance*).

(*c*) The effects of different treatments or conditions are additive (*additivity*).

(*d*) Measurement is on an interval or ratio scale (see 9.C.7).

In addition, most parametric tests require that associations between variables are linear (the assumption of *linearity*).

Behavioural data frequently violate some or all of these assumptions, thereby undermining the validity of parametric methods in many cases. For example, behavioural data sometimes exhibit a highly skewed distribution, violating the assumption that data are normally distributed (see 9.H on data transformations). The historical reason for this disparity between the ideal world of parametric statistics and the reality of behavioural data is probably that parametric methods were originally developed for analysing agricultural data. Here, the sample units were plots of land, the outcome measures were the quantity and quality of crop yields and the various assumptions of parametric tests were usually realistic.

2 *Nonparametric tests,* such as the Mann-Whitney *U* test, Wilcoxon matched-pairs test, chi square and Spearman rank correlation, are generally slightly less powerful than the equivalent parametric tests. However, because they are free from the assumptions of parametric tests they are more **robust**; that is, less dependent on various assumptions about normality and so on for their validity. Moreover, since nonparametric tests require only ranks, rather than measurements on an interval or ratio scale, they can be used to analyse data measured on an ordinal scale. (The Wilcoxon matched-pairs test is something of an exception, in that it requires ordinal measurement not only for each score, but also for the differences between pairs of scores.) Some nonparametric tests (such as chi square) can be used to analyse data measured on a nominal scale.

The statistical power (see 9.C.4) of many nonparametric tests is almost as great as that of the equivalent parametric test. For example, under conditions where parametric tests might properly be used, the Wilcoxon matched-pairs test and the Mann-Whitney U test are 95% as powerful as the most powerful equivalent parametric test (the t test). Similarly, Spearman rank correlation is about 91% as powerful as Pearson correlation (Siegel, 1956). Moreover, by increasing the sample size a nonparametric test can always be made as powerful as the equivalent parametric test. Therefore, the robustness of nonparametric tests need not be bought at the expense of greatly reduced statistical power. In any case, if a nonparametric test is powerful enough to detect a significant effect, then it is powerful enough – or, as one statistician put it, 'If the tree falls, the axe was sharp enough'.

In addition to their relative freedom from unrealistic assumptions, nonparametric methods have two practical advantages over the equivalent parametric methods. First, they are generally easier to calculate by hand without recourse to a computer – an advantage that continues to shrink as computers become ever cheaper. Second, they are better suited to working with small ($n < 10$) sample sizes – a situation that behavioural scientists regrettably face all too often.

We should add that nonparametric tests are not entirely assumption-free: they still assume that measurements are independent (see 2.E) and most assume that variables have underlying continuity. However, these assumptions are usually realistic for behavioural data.

In conclusion, then, nonparametric methods are frequently the tools of choice for the analysis of behavioural data. However, for some more complicated types of analysis (for example, certain types of multivariate analysis), only parametric techniques are available (although see Conover, 1980 and Meddis, 1984 for recent developments in nonparametric multivariate analysis).

G The uses and abuses of correlations

Correlations are easily calculated, seemingly easy to understand and widely used in the analysis of behavioural data. Unfortunately, correlations are also easy to misuse and misinterpret. First, a brief reminder of what a correlation is:

1 *The meaning of a correlation.* A correlation coefficient describes the extent to which two measures (or variables) are associated, or vary together. Two measures (for instance, height and weight) are positively correlated if high scores on one measure are associated with high scores on the other measure, and low scores on the first are associated with low scores on the second. Conversely, if high scores on one measure are associated with low scores on the second, and vice versa, then the two measures are negatively correlated.

The strength of the association is indicated by the size of the correlation coefficient, which is a number between -1.0 and $+1.0$. A correlation of ± 1.0 indicates a perfect association (i.e., every score on one measure is perfectly predicted by the scores on the other measure), while a correlation of 0 means there is no linear association between the two measures (i.e., knowing one set of scores provides no information about the other set). A *significant* correlation is generally taken to mean a correlation that differs significantly from zero.

For an illustration of the strength of association represented by correlations of different sizes, see Fig. 9.1. Note that at first sight a weak but statistically significant correlation (Fig. 9.1*d*) can appear little different from a complete absence of correlation (Fig. 9.1*c*; see 9.D on effect size versus significance).

2 *Pearson versus Spearman correlations.* The two most commonly used correlation coefficients are the **Pearson** (or product moment) correlation coefficient (r) and the **Spearman** rank correlation coefficient (r_s). A Pearson correlation (sometimes referred to as just 'correlation') is a parametric statistic (see 9.F), and therefore requires

Fig. 9.1. Four scatter plots, illustrating the association between two variables represented by various correlations. All correlations are Pearson correlation coefficients, with 38 degrees of freedom ($n = 40$). All significance levels are two-tailed.
(a) $r = +0.88$ ($p < 0.001$); (b) $r = -0.85$ ($p < 0.001$); (c) $r = +0.03$ (n.s.; $p > 0.10$); (d) $r = +0.40$ ($p < 0.02$).
Note that although the correlation shown in (d) represents a relatively weak association, it is, nonetheless, statistically significant.

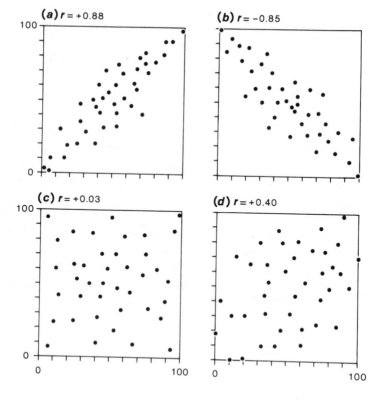

that both variables are measured on an interval or ratio scale; both variables are normally distributed; and the variations in scores of both variables are roughly equal (although the Pearson correlation is reasonably robust when there is departure from normality; see Sprinthall, 1982, p. 193).

The Spearman rank correlation is a nonparametric statistic, requiring measurement on an ordinal scale or higher. The discussion which follows applies to both types of correlation coefficient, which are usually similar in size when calculated for the same data. (Incidentally, an electronic calculator which gives Pearson correlations can be used to calculate Spearman correlations, by calculating the Pearson correlation for the ranks rather than the raw data.)

3 *Correlation does not imply causation.* A statistical correlation between two variables does not mean they are directly related in a causal manner. In other words, a correlation between *A* and *B* does not by itself demonstrate that *A causes B* (or that *B* causes *A*). For example, if two church clocks happen to be set so that one clock always strikes a few seconds before the other, then an observer would detect a strong temporal correlation between the striking of the two clocks. It would obviously be wrong, though, to infer that the striking of the first clock *causes* the striking of the second (Cullen, 1972). Similarly, Sprinthall (1982, p. 210) quotes a study which found that the amount of time teachers spend smiling is positively correlated with the results achieved by their students. However, we cannot infer from this that smiling causes students to achieve better results. The reverse might be true: perhaps good results cause teachers to smile more. Another possibility is that teachers who smile a lot also tend to award higher grades because their standards are different. The point is that we cannot distinguish between these various causal hypotheses from the correlational evidence alone. To do this would require an experiment in which one variable (say, the amount of smiling by the teacher) was systematically varied.

A correlation between two variables, *A* and *B*, can arise for one of three reasons: *A* causes *B*; *B* causes *A*; or *A* and *B* are independently related to a third variable, *C*. Problems frequently arise with this last case; that is, when the two measured variables are independently associated with an unknown third variable. For example, the population of the world and the age of the Queen of England are positively correlated – not because they are directly linked by a causal relationship but because they are both correlated with a third variable: time.

4 *Correlations refer to linear relations between two variables.* A Pearson correlation is meaningless if the two measures are associated in a non-linear manner. Specifically, a lack of correlation between two measures does not imply a lack of any association between them if they are non-linearly related. Fig. 9.2 illustrates the case of two variables (*A* and *B*) which are strongly associated, but according to an inverse *U*-shaped relation rather than a linear relation. The correlation between *A* and *B* is zero, but it would obviously be wrong to infer that there is no association between them. Thus, it is essential to verify that two variables are not associated according to some non-linear relation before calculating the correlation between them.

Fig. 9.2. Scatter plot showing an inverse U-shaped relation between two variables (*A* and *B*). The association between *A* and *B* is strong, but non-linear. However, the Pearson correlation between them is zero ($r = 0.0$). Clearly, the lack of correlation does not imply a lack of any association between the two variables.

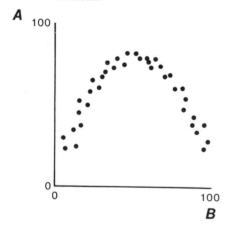

The best precaution is always to *plot the data on a scatter plot* (as shown in Figs. 9.1 and 9.2) before calculating correlations. (Strictly speaking, Spearman rank correlation requires only that the relation between the two measures is monotonic increasing or decreasing rather than specifically linear.)

5 *Correlation coefficients should not be averaged.* Correlation coefficients are not like ordinary numbers and do not obey the normal rules of arithmetic. Therefore, it is incorrect to average several correlation coefficients by calculating their arithmetic mean. (However, the difference between two Pearson correlations is meaningful and its statistical significance can be tested, although there is no way of testing the significance of a difference between two Spearman correlations.)

To find the average of several Pearson correlations, each correlation should first be converted to the corresponding **Fisher z transform**, where $z = \frac{1}{2} \ln [(1+r)/(1-r)] = \tanh^{-1} r$. These z transforms are then averaged by calculating their arithmetic mean. Finally, the mean z transform is converted back into a correlation using the inverse transformation, and it is this value of r that represents the true average of the correlations. Tables for transforming r into z and vice versa can be found in many statistics books (e.g., Snedecor & Cochran, 1980). The best way of averaging several Spearman rank correlations is simply to calculate their median value.

A related point is that two correlations cannot be directly compared as a ratio in the simple way that, say, two weights or two distances can be compared. For example, a correlation of 0.8 does not represent an association that is twice as strong as a correlation of 0.4. One way of comparing Pearson correlations is by using the square of the correlation coefficient (r^2), which is known as the **coefficient of determination**. Broadly speaking, r^2 is the proportion of the variation in one measure that is accounted for statistically by the variation in the other measure. Hence, a correlation of 0.8 means that 64% of the variation in one set of scores is accounted for statistically by the variation in the other ($r^2 = 0.64$). With a correlation of 0.4, one measure accounts for only 16% of the variation in the other measure ($r^2 = 0.16$). In this sense, a correlation of 0.8 is *four* times as great as a correlation of 0.4.

6 *A correlation can be misleading if the underlying population is not homogeneous.* When interpreting a correlation it is usual to assume that the strength of association between the two variables is the same for all values of both variables; in other words, that the underlying population is homogeneous. Sometimes, though, this assumption is not justified because the strength of association is different for different members of the population. For example, two variables (*A* and *B*) may show no association over most of their range, yet they can still be highly correlated if a strong association exists for extreme values only. This possibility is illustrated in Fig. 9.3. The danger with this type of correlation is that it falsely implies a general association between two variables when in fact the association is restricted to extreme cases. The association is, therefore, of limited validity.

Fig. 9.3. Scatter plot showing two variables (*A* and *B*) which are positively correlated only for extreme values of *A* and *B*. The overall correlation is high (Pearson $r = +0.71$, 30 d.f.; $p < 0.001$, two-tailed), yet the association between *A* and *B* is very weak for most values of *A* and *B*.

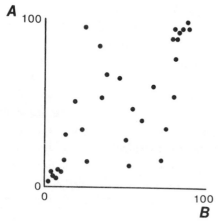

Similarly, a low correlation can arise if two variables are positively correlated for one sub-group of the population (say, males) but negatively correlated for another (say, females). This possibility is illustrated in Fig. 9.4. In this case, the low overall correlation that results from not separating out the two distinct sub-groups is actually unrepresentative of all individuals in the population. Once again, the best precaution is to inspect the results on a scatter plot before calculating the correlation. The presence of distinct sub-groups or 'outliers' within a sample will usually be apparent from a scatter plot.

Fig. 9.4. Scatter plot showing the association between two variables (A and B). The overall correlation, calculated for the whole sample, is close to zero (Pearson $r = +0.08$, 30 d.f.; $p > 0.10$, two-tailed). However, the sample is composed of two distinct sub-groups. For one sub-group (say, males; indicated by filled circles) there is a very strong positive correlation between A and B ($r = +0.92$, 14 d.f.; $p < 0.001$, two-tailed), while for the other sub-group (say, females; indicated by open circles) there is a strong negative correlation ($r = -0.89$, 14 d.f.; $p < 0.001$, two-tailed). It would be misleading only to calculate the correlation for the sample as a whole, since the underlying population is not homogeneous.

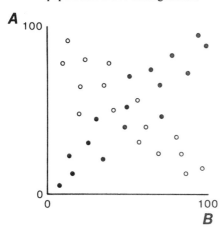

7 *The statistical significance of a correlation says nothing about the size or importance of the correlation.* The distinction between effect size and statistical significance has already been discussed in 9.D, but is important enough to warrant repetition here. A tiny correlation can be highly statistically significant provided the sample size is large enough. For example, with a sample size of 100 a correlation of 0.20 is statistically significant ($p < 0.05$, two-tailed), even though a correlation of this size represents an extremely weak association ($r^2 = 0.04$; i.e., only 4% of the variation in one measure is accounted for by variation in the other).

We suggest that it is worth remembering the following informal interpretations for statistically significant Pearson correlations of various sizes (quoted from Sprinthall, 1982, table 9.3):

Value of r	*Informal interpretation*
< 0.2	Slight; almost negligible relationship
0.2–0.4	Low correlation; definite but small relationship
0.4–0.7	Moderate correlation; substantial relationship
0.7–0.9	High correlation; marked relationship
0.9–1.0	Very high correlation; very dependable relationship

These verbal tags, which were originally suggested by the statistician Guilford, are essentially arbitrary and apply only to statistically significant correlations. Their value lies in emphasising the point that statistically significant correlations may represent associations that are so weak as to be negligible.

8 *'Fishing expeditions' are a mixed blessing.* We are referring here to the practice of calculating correlations for every possible pairwise combination of variables, or 'correlating everything with everything'. This is often deprecatingly referred to as 'fishing' when it stems from a failure to formulate clear hypotheses before the data were collected. The implication is that the researcher must resort to analysing the data in every possible way in order to dredge up at least some significant results, which are then explained by a *post hoc* 'hypothesis'.

The danger of 'fishing' for correlations in this way is that if a very large number of correlation coefficients are calculated, some statistically significant correlations may arise by chance. Therefore, any conclusions drawn from such correlational analysis should take into account the *total* number of correlations that were calculated (significant and non-significant). A significant correlation may seem less compelling if it is the only one out of, say, 45 correlations originally calculated.

The second point rests on the distinction made earlier (9.B) between exploratory and confirmatory data analysis. Systematically searching through data for interesting results is entirely admissible – indeed, highly recommended – provided it is seen solely as *exploratory* analysis, and not as confirmatory analysis or 'hypothesis-testing'. Using a significant correlation as empirical proof of the *post hoc* 'hypothesis' which it gave rise to in the first place is, clearly, tantamount to placing a bet after the horse has won the race.

H Data transformations

As we pointed out earlier (9.F), parametric statistical tests require various assumptions to be made about the nature of the data, but these assumptions are frequently violated by behavioural data. One assumption is that the data are normally distributed. In practice, the consequences of moderate departures from a normal distribution are not always too serious. For example, Student's *t* test is relatively insensitive to deviations from a normal distribution, except in the case of an independent-samples design with unequal sample sizes. (The validity of parametric tests does, however, depend strongly on other assumptions, such as homogeneity of variance and additivity.)

Data which are not normally distributed – for instance, data following a skewed distribution – can sometimes be transformed so that they become approximately normally distributed. The guidelines for transforming data are as follows:

1 *Square root transformation.* Data in the form of *counts*, for example, true frequencies or total numbers of occurrences, are likely to follow a Poisson rather than normal distribution. This type of non-normality can often be corrected using the square root transformation (i.e., convert x to $\sqrt{x}$). If the data include zero scores, then add 0.5 to all values (i.e., convert x to $\sqrt{(x + 0.5)}$).

2 *Arcsine–square root transformation*. When data are in the form of *proportions* or *percentages* (for example, when time sampling is used) non-normality can often be cured by transforming data according to the arcsine-square root (or angular) transformation; i.e., convert each value to arcsine($\sqrt{p}$), where p is a proportion ($0 < p < 1$). If the scores fall between 0.3 and 0.7, then it is not usually necessary to use the transformation.

3 *Logarithmic transformation.* The logarithmic transformation (convert x to $\log(x)$) is useful in a number of cases; for example, when the mean and variance are positively correlated or when the frequency distribution is skewed to the right. If the data include zero scores, add 1 to each score (i.e., convert x to $\log(x + 1)$).

As we have stressed, none of the common parametric tests depend solely on the normality of data, so normalising data does not by itself ensure their validity. However, transforming data sometimes cures more than one type of departure from the assumptions of parametric tests; for example, by making data fit the assumptions of homogeneity of variance and additivity as well as normality. For a discussion of transformations, see Sokal & Rohlf (1981).

I Measurement error versus sampling error

1 *Measurement error* is the combined error that results from inevitable imperfections and variability in the process of measurement. Measurement error may be random (scores equally likely to be greater or less than their true values) or systematic (scores consistently either greater or less than their true values).

2 *Sampling error* is the difference between a sample statistic (such as the sample mean, $\overline{X}$) and the population parameter that is being estimated (such as the population mean, μ). Sampling error arises because a finite sample rather than the whole population is measured, and can occur in any study (unless the entire population is measured).

If a population is sampled several times the various sample means will, on average, be evenly distributed about the population mean. In statistical terms, each sample mean is an unbiased estimate of the population mean. The standard deviation of the sample means is referred to as the **standard error of the mean** (SEM), and is estimated by dividing the standard deviation (σ) by the square root of the sample size (n); i.e., SEM $= \sigma/\sqrt{n}$. Note that as the sample size (n) increases, the standard error of the mean gets smaller, showing that larger samples give more certain estimates of the true (population) mean. Note also that the standard error of the mean is inversely related to the square root of the sample size. Thus, increasing the sample size has a diminishing effect on improving the estimate of the population mean.

J   Statistics books

Nonparametric Statistics for the Behavioral Sciences by Siegel (1956) is still the bible of nonparametric statistics, despite its age. It is tried, trusted, widely used and written with admirable clarity. Conover (1980) and Meddis (1984) give more recent accounts of nonparametric statistics, including many tests developed since Siegel's book was written. Among the best introductory statistics books, dealing mainly with parametric methods and offering clear accounts of the basic concepts, are Bailey (1981), Folks (1981) and Sprinthall (1982). Of the more advanced texts, we recommend Snedecor & Cochran (1980), Sokal & Rohlf (1981) and Zar (1984). Exploratory data analysis is presented in detail by Tukey (1977) and at a simpler level by Velleman & Hoaglin (1981). For an introduction to multivariate analysis – methods which can deal with many variables simultaneously, such as principal component analysis, factor analysis, multiple linear regression and cluster analysis – see Maxwell (1977) and chapters in Colgan (1978). For an alternative view on the logic of statistical inference and an introduction to the concept of likelihood, see Edwards (1972). Batschelet's (1979) book on basic mathematics for biologists is also useful (see also Newby, 1980).

K **Summary of common statistical tests** (after Siegel, 1956 and Conover, 1980)

Most common statistical tests are tabulated here, according to the purposes for which they are used. All tests, except those marked with an asterisk (*), are nonparametric and require measurement at the ordinal level or greater unless otherwise specified.

1. Does the sample come from a specified population? (Tests of goodness of fit for single samples)
- CHI SQUARE test for one sample (nominal data)
- BINOMIAL TEST (nominal data)
- KOLMOGOROV-SMIRNOV one-sample test

2. Is there a significant difference between the scores of two unrelated samples; for example, between the scores of two different groups of subjects? (Tests of difference between two unmatched samples)
- CHI SQUARE test for two independent samples (nominal data)
- FISHER exact probability test (nominal data)
- MANN-WHITNEY *U* test
- STUDENT'S *t* TEST for unmatched samples ('*t* test')*

3. Is there a significant difference between the scores of two related samples; for example, between the scores of the same subjects under two different conditions, or between siblings? (Tests of difference between two matched samples)
- WILCOXON matched-pairs signed ranks test
- STUDENT'S *t* TEST for matched samples ('matched-pairs *t* test')*

4. Are there significant differences between the scores of several unrelated samples? (Tests of difference between *k* unmatched samples)
- CHI SQUARE test for *k* independent samples (nominal data)

- KRUSKALL-WALLIS one-way analysis of variance
- ANALYSIS OF VARIANCE*

5 Are there significant differences between the scores of several related samples; for example, between the scores of the same subjects measured under several different conditions? (Tests of difference between k matched samples)
- FRIEDMAN two-way analysis of variance
- REPEATED MEASURES ANALYSIS OF VARIANCE*

6 Are two sets of scores associated? (Measures of correlation between two samples)
- SPEARMAN rank correlation coefficient
- KENDALL rank correlation coefficient
- PEARSON product-moment correlation coefficient ('correlation')*

7 Are several sets of scores associated; for example, are the scores of one group of subjects consistent when measured several times, or is there an overall association between several different measures for the same set of subjects? (Test of concordance between k rankings of the same subjects)
- KENDALL COEFFICIENT OF CONCORDANCE

10

Miscellaneous issues

A Tests of preference and differential responsiveness

Tests of whether a subject responds more strongly to one stimulus rather than another may be conducted and analysed in a variety of ways, both in the field and in the laboratory. Several important issues arise when considering choice tests: (*a*) Should the stimuli be living animals, stuffed animals, audio or visual recordings, or models? (*b*) Should the stimuli be presented simultaneously or successively? (*c*) Should the same subject be tested once or repeatedly? (*d*) Should the types or rates of response to the stimuli be compared in terms of absolute or relative differences? Each method has its own advantages and drawbacks and these need to be considered carefully before proceeding with a study.

1 *Stimuli.* Living animals are the richest and most natural type of choice stimulus, but they have several potential drawbacks. First, the stimulus animals are liable to interact with the test subject, making it difficult to decide who chooses whom. Sometimes these interactions can be eliminated if the stimulus animals are made unaware of the subject's presence by using one-way screens (see 2.B). Even if one-way screens are used, though, the behaviour of the stimulus animals is likely to change with successive tests as they become habituated. If, on the other hand, different individuals are used as models in successive tests they may behave differently from one another. Finally, because a living animal does provide such a rich and complex stimulus, it can be difficult to know precisely which features of the stimulus influenced the subject's choice.

Stuffed animals and recordings have many of the advantages of living stimuli and, in addition, they are not influenced by the subject. Furthermore, the effectiveness of particular features of the stimuli can be analysed by selective removal or alteration of different parts of the choice stimuli. This type of manipulation is particularly easy with audio and video tape-recordings. However, the responses of the test animal might be greatly reduced if presented with a stuffed stimulus animal rather than a real one.

Models can be surprisingly effective as test stimuli. They are relatively simple to make and, since their characteristics can be varied systematically, the results obtained with them are easily analysed. However, responses to models are often greatly reduced compared with those to more natural stimuli.

2 Simultaneous or successive tests? In principle, comparison between stimuli presented at the same time provides the more sensitive measure of preference, while presentation of each stimulus in turn provides the more rigorous test. However, drawbacks with both methods of testing need to be considered.

Simultaneous presentation may be distracting to the subject and also tends to be highly unnatural. The subject may become 'trapped' by its first choice, simply because it happens to be facing one way and approaches the stimulus it is facing. Finally, withdrawing from one stimulus may be incorrectly interpreted as approaching the other (or vice versa). This last problem can sometimes be overcome by providing a three-way choice, with a 'blank' (no-stimulus) choice in addition to the two choice stimuli.

Successive presentation may be insensitive if the subject responds at the maximum possible level to each of several stimuli when it cannot make simultaneous comparisons between them. More seriously, successive tests may be confounded by order effects because the subject becomes generally more (or less) responsive to all stimuli during the course of testing. Such effects may be allowed for by varying the order of presentation for different subjects, or by presenting two stimuli (A and B) in the order A,B,B,A. Alternatively, a Latin Square experimental design (see Lehner, 1979, p.81, or Clarke, 1980, p.119) can be used to eliminate order of presentation effects.

3 *Repeated testing?* On the face of it, the more information that a given subject provides the better. Testing the subject many times is likely to enhance the reliability of the results. However, the subject may simply stop responding if it is tested too many times and, more seriously, its preferences may be influenced by repeated exposure to the stimuli. This effect is seen, for example, in imprinting experiments with young precocial birds, when the bird starts to a form a preference for the novel stimulus with which the familiar stimulus is being compared (e.g., Bateson, 1979*b*). Also remember that all the data provided by one subject must only be used as one member of the sample in subsequent statistical analysis, since, in this context, repeated tests with the same subject must not be treated as though they were statistically independent (see 2.E). For this reason, increasing the time spent testing one subject eventually has diminishing returns. The actual time allotted to each subject will obviously depend, among other things, on the availability of subjects.

4 *Analysing and presenting data.* If each subject's reaction to a stimulus is recorded as an all-or-nothing response (for example, 'approaches' or 'does not approach' the stimulus), the simplest method of presenting the results is in terms of how many subjects responded to each stimulus. However, this measure of response is relatively crude. If quantitative measures of how much or how many times each subject responded to the stimuli were obtained (which they should have been if at all possible), forms of analysis that use this additional information are preferable.

Absolute differences in responsiveness are obtained by subtracting the response to stimulus B from the response to stimulus A. This provides a single score for each subject. If the same subject has been tested repeatedly, the mean of the differences in response to the two stimuli should be used as that subject's score. The scores for all subjects can then be analysed statistically by means of a matched-pairs test, such as a Wilcoxon signed-ranks test or a matched-pairs *t* test (see 9.K).

Response ratios can be calculated in a variety of ways, a particularly useful form being:

Responses to A /(Responses to A + Responses to B)

If the subject has always responded to A and never to B, its score will be 1.0. Conversely, if it has responded to B but not A, its score will be 0. The chance level of response is 0.5. If more than two stimuli have been used, the divisor is the total number of responses to all the stimuli and the chance score is 1.0 divided by the number of stimuli used (for example, 0.25 in the case of four stimuli) .

Subjects which fail to respond to either stimulus (*non-responders*) can cause problems in the analysis stage (particularly if this is done on a computer), since calculating their scores involves dividing by zero. This problem can be dealt with by either (*a*) systematically excluding the scores of all non-responders, or (*b*) arbitrarily assigning a chance-level score (0.5 for a two-choice test) to all non-responders.

For the purposes of statistical analysis, each subject provides one value, which is the difference between its ratio score and the chance level. These difference scores are analysed with a matched-pairs test, as in the case of absolute differences. It is worth noting that, for various reasons, scores obtained by the ratio method tend to be unevenly distributed, with modes at either end of the distribution between 0 and 1.0. This non-normal distribution of scores may render invalid certain types of statistical analysis (see 9.F).

Results based on absolute rather than ratio scores will be greatly affected by those individuals which have responded strongly in the tests. By contrast, the ratio method is more sensitive to variation within individuals, even when considerable variation is found between individuals. Indeed, it is logically possible for the two methods to generate contradictory results if the high-responding individuals have a preference for stimulus A and the low-responders a preference for B. For this reason, it is wise to analyse results both ways and, if the two methods generate different conclusions, consider carefully why the differences have arisen. The apparent contradiction may itself be interesting.

B Measuring bout length

Behaviour patterns sometimes occur in temporal clusters, referred to as **bouts**, in which the same, relatively brief act is repeated several times in succession (a bout of *events*; see 3.F) or the same, relatively prolonged behaviour pattern occurs continuously for a period (a bout of a single behavioural *state*). For example, the pecking behaviour of domestic chicks occurs in discrete bouts, each bout consisting of many, rapidly occurring pecks (events) with successive bouts separated by fairly long gaps (e.g., Machlis, 1977).

If behaviour patterns are neatly clumped into discrete bouts separated by uninterrupted gaps, then it is relatively easy to distinguish one bout from the next. Often, though, bouts are not obviously discrete, in which case a statistical criterion must be used to define a single bout of behaviour. One commonly used technique is **log survivorship analysis**. This is a simple graphical method for specifying objectively the minimum interval separating successive bouts: the **bout criterion interval** (**BCI**). Any gap between successive occurrences of the behaviour that is less than BCI in length is treated as a *within-bout* interval, while all gaps greater than BCI are treated as *between-bout* intervals.

The method is based on the assumption that within- and between-bout intervals are generated by two random (Poisson) processes with different rate constants, giving short within-bout intervals and long between-bout intervals. The slope of the log survivor function is proportional to the probability that the behaviour will occur again within the corresponding interval. A random distribution of intervals between successive occurrences will generate a linear curve on a log survivorship plot. If occurrences are clumped – with many, short within-bout intervals and a few, long between-bout intervals – the log survivorship curve will have two parts and the point where these join marks an objective criterion for deciding the bout length (BCI).

To estimate BCI, the cumulative frequency of gap lengths (on a logarithmic scale) is plotted against gap length (on a linear scale). Fig. 10.1 shows a stylised example of a log survivorship plot. If the assumptions of the statistical model are met, the graph should have two fairly distinct parts: a rapidly declining portion representing the short within-bout gaps, and a slowly declining portion representing the long between-bout gaps. A rough value for BCI is given by estimating visually the point at which the slope of the graph changes most rapidly. The interval corresponding to this 'break point', if it can be identified, is usually taken as an objective estimate of BCI (e.g., Slater, 1974). However, Slater & Lester (1982) have pointed out that this simple method does not give the best possible estimate of BCI in terms of minimising the number of intervals that are wrongly classified as between- or within-bout intervals. The best criterion, for which they give a formula, is actually a slightly longer interval than the one corresponding to the break point.

Fig. 10.1. The general form of a log survivorship plot, used for determining bout length. The time interval between successive occurrences of the behaviour pattern is t. Log (N) is the logarithm of the number of intervals greater than the corresponding value of t. BCI is the *bout criterion interval*, an objective estimate of the minimum interval that distinguishes separate bouts (i.e., the maximum within-bout interval). (After Slater & Lester, 1982.)

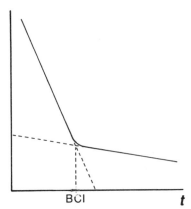

The smaller the difference in slope between the two parts of the log survivor function, the more intervals will be wrongly classified. If the slope changes gradually and there is no clear break point on the graph then it is probably wrong to split the behaviour into bouts anyway. In general, if the presence of bouts is not reasonably obvious, there seems little point in using sophisticated statistical techniques to split the data into bouts.

C **Plotting data**

As we argued earlier, a graph, histogram or scatter plot is generally much more informative than tables of numbers. More importantly, plotting data before analysing them statistically is a wise precaution against misinterpreting the statistics (see, for example, 9.G on the need for drawing scatter plots before calculating correlations). We suggest the following general guidelines for plotting data.

1 *Each axis should be clearly labelled* with the name of the variable (e.g., average distance travelled, vocalisation rate or amount of food consumed) and the units in which the variable is measured (e.g., $km\ d^{-1}$, min^{-1} or $g\ h^{-1}$). The type of measure used for behavioural categories should be clearly indicated: reciprocal units of time (s^{-1}, min^{-1}, d^{-1}, etc.) for frequency measures; and units of time (s, min, d, etc.) for duration and latency measures. The proportion of time spent performing an activity and time sampling measures (4.D, 4.E) are expressed as dimensionless scores between 0.0 and 1.0. Where a measure is expressed as the total number of occurrences, the period of time over which the events were counted should be indicated (e.g., 'per 30 min').

Results are often presented in the form of multiple plots, all sharing the same x-axis but each with a different y-axis. This type of multiple plot can be misleading when casually inspected unless attention is drawn to the fact that the y-axis scales differ. It is doubly important in such cases that the axes are clearly labelled.

2 *Bear in mind that graphs using a restricted range for the y-axis can be misleading.* The practice of using a restricted range along the y-axis, starting from a non-zero value, means that small variations in the dependent variable (y) are accentuated. It may be necessary and justifiable to focus on a particular part of the y-axis range, as though using a magnifying glass to inspect the data, since it can be very useful in revealing patterns in the data. However, unless it is made perfectly clear that the y-axis shows only a selected portion of the range, the graph can be highly misleading. Be cautious, therefore, when interpreting graphs that show only selected portions of the range over which the variables were actually measured.

3 *Measures of variability.* When sample means (as opposed to individual scores) are plotted on a graph or histogram, it is often helpful to show the standard errors of these estimates as well. These are shown by drawing a vertical bar through each point, the length of the bar representing the size of the standard error.

It is common practice in the behavioural literature to draw an error bar proportional in length to two 'standard errors' (i.e., ± 1 standard error of the mean or SEM; see 9.I.2). A frequent mistake is to assume that if the error bars of two means do not overlap, then the means must be significantly different. This is not necessarily true if the error bars represent ± 1 SEM. A more conservative index of variability would be ± 2 SEM (or ± 1.96 SEM, to be precise), since 95% of all estimates of the mean (sample means) should fall within ± 1.96 SEM of the true (population) mean. Variability of scores about a median is often indicated with vertical bars denoting the inter-quartile range of the scores.

D Ratings of individual distinctiveness

Many people who spend a long time watching the same animals come to feel that individuals have distinctive personalities. These impressions may lend colour to qualitative accounts of the animals' behaviour, but are not normally regarded as especially useful in quantitative studies. Nevertheless, such impressions can be reliably and validly rated. Data derived from observers' evaluations of individuals are used routinely in studies of human personality (Cairns & Green, 1979), and terms such as 'personality' or 'individual distinctiveness' are undoubtedly useful in referring to behavioural characteristics that reliably differentiate one individual from another.

A method of rating observers' evaluations, comparable to those used in studies of humans, has been applied to rhesus monkeys (Stevenson-Hinde *et al.*, 1980) and domestic cats (Feaver *et al.*, 1986). The technique is generally reliable and can, to some extent, be validated against other forms of measurement (see 6.A). It may provide a useful means of measuring aspects of individual differences that are not easily assessed in other ways, because the ratings are of overall patterns of behaviour rather than discrete events.

The method uses forms listing all the individual animals which are to be rated, with a linear scale drawn for each individual. One form is used for each category, or **item**, on which the individuals are to be rated. Each item is an adjective (such as *active, curious, aggressive, playful*) which has a behavioural definition. A cross is marked on each scale at a position corresponding to the observer's overall assessment of how much that individual expresses that behavioural item (Fig. 10.2). Subsequently, the distance from the left-hand end of the line to the position of each cross is measured. This distance is used as the numerical score for that individual on that item. The left-hand end of the scale represents the minimum and the right-hand end the maximum expression of the behavioural item among *all* the individuals being rated. Even though an individual may show an extreme expression of an item every now and then, it is not marked as being at the end of the scale if it is generally less extreme on this item.

Needless to say, the method requires the human rater to be thoroughly familiar with the individual animals. It also requires careful definition of each behavioural item before the individuals are formally assessed (see 3.D).

The following are some of the definitions that have been used for rating rhesus monkeys: *Effective*, gets own way, can control others; *Aggressive*, causes harm or potential harm; *Excitable*, over-reacts to change; *Sociable*, seeks companionship of others; *Playful*, initiates play, joins in when play is solicited. All of these items have been validated by showing that they are significantly correlated with direct observations of the monkeys' behaviour (Stevenson-Hinde, 1983). However, since ratings do not measure the same things as direct observation, it is not possible to make direct comparisons for all items.

Observers' ratings do not replace direct recording methods and, in many cases, it would be quite inappropriate to use them in a study of individual variation. Nevertheless, they can provide useful information about subtle aspects of an individual's style that is not easily obtained in other ways. Direct observations of behaviour are in some respects similar to measuring a human face with a ruler.

Fig. 10.2. An example of the type of coding form used to record observers' ratings of individual distinctiveness. One form is used for each category, or *item*. Each line corresponds to an individual subject ('Ari...', 'Don...', 'Elf...', etc.). The left-hand end of the scale represents the minimum expression, and the right-hand end the maximum expression, of that item for all individuals in the sample studied. The rater marks a cross on the scale, corresponding to the assessment for that individual on that item. This is subsequently converted to a numerical score by measuring the distance (in mm) of the cross from the left-hand end of the scale. (After Feaver *et al.*, 1986.)

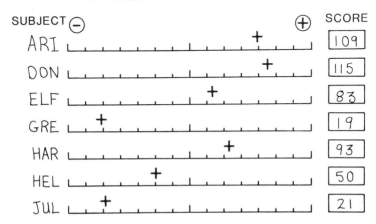

CATEGORY : Active

SUBJECT ⊖	⊕	SCORE
ARI	+	109
DON	+	115
ELF	+	83
GRE	+	19
HAR	+	93
HEL	+	50
JUL	+	21

Each measurement can be precise and consistent, and yet uncovering an individual's distinctiveness from combinations of the measurements is virtually impossible – at least with present-day techniques. The rating method provides a higher level of description, because it can capture the overall pattern of an individual's behaviour that remains elusive when discrete events are measured. The overall pattern involves behaviour occurring in a wide variety of conditions and it may also take into account what happens in complex social interactions. Direct recording is usually immediate, being done at the time each behaviour pattern occurs, whereas an observer's ratings of an individual's style are formed over a reasonably long period of time. The usefulness of the rating method lies precisely in this difference, because the human rater has played an active role in filtering, accumulating, weighting and integrating information over a considerable period of time (Block, 1977). Moreover, ratings of general attributes may be more predictive over time than specific measures of behaviour, because the meaning and significance of a particular behaviour pattern may change with development.

We have deliberately avoided applying the term 'subjective' to this method of assessment, since it falsely implies that observational methods are, in contrast, 'objective'. Even direct observational recording requires the observer to make judgements and interpretations about whether or not the subject is performing a particular behaviour pattern (Cairns & Green, 1979).

It is, of course, especially important that observers using such ratings should check that they are reliable and, where possible, validated against other forms of measurement (see 6.A). If reliability or validity are measured using correlations, these correlations should be large (say, $r > 0.7$), as well as statistically significant (see 9.D on the distinction between effect size and statistical significance).

E The ethics of animal research

The ethical implications of causing suffering and disruption to animals in the course of scientific research is an important issue which everyone studying animals should be aware of. Many would argue that those who study animal behaviour but are insensitive to the condition and welfare of their animals lay themselves open to the charge that their science is worthless. In recent years, public attitudes towards the use of animals in research have changed markedly and are reflected in the increased awareness of many scientists. In order to tackle some of these difficult ethical problems, the Association for the Study of Animal Behaviour (ASAB) in the UK and the Animal Behavior Society (ABS) in America have formed, respectively, an Ethical Committee and an Animal Care Committee. The quotations in this section are taken from their joint Guidelines for the Use of Animals in Research (ASAB/ABS, 1981), which are reproduced in full in Appendix 1. Many of the points raised here apply to observational and field studies as well as laboratory-based experiments.

The ethical issues of using animals in research are complex and require at least two difficult types of judgement to be made. First, how worth while, both scientifically and in terms of possible practical benefits, is a proposed piece of research? Second, how much suffering to animals is likely to result from the research? We agree with the view that 'While the furthering of scientific knowledge is a proper aim, and may itself advance an awareness of human responsibility towards animal life, the investigator should always weigh any potential gain in knowledge against the adverse consequences for the animals used as subjects, and also for other animals in the ecosystem in the case of field studies.' This point is represented in Fig. 10.3. The hatched area indicates categories of research that ought not to be carried out on ethical grounds, either because they involve an unacceptable amount of animal suffering or because they are likely to be of poor quality scientifically. Needless to say, assessing the quality of a proposed research project or specifying what is meant by 'high-quality' research can be difficult. 'Quality' is a multidimensional issue which includes both scientific excellence (that is, whether the research is likely to make an important contribution to understanding) and potential practical benefits (which may be short-term or long-term). Nonetheless, a similar problem is regularly tackled by the many committees which dispense money to finance scientific research and by editorial boards of scientific journals. Few would claim that such judgements are ever precise, but a consensus can often be reached when assigning research proposals to one of a few, broad classes. A similar process could, in principle, be applied to judging the amount of animal suffering that is likely to be caused.

Ethical issues do not apply solely to laboratory experiments, since 'Even observation of free-living animals in their natural habitat may involve disruption, particularly if feeding, trapping or marking is involved.' Simply observing wild animals can have marked effects on them (see 2.B), while capturing and marking them for identification can be stressful and disruptive. Indeed, the scientifically desirable practice of habituating wild animals to the presence of an observer

Fig. 10.3. A representation of how the amount of animal suffering involved in a proposed research project can be balanced against the quality of the research. The hatched area represents research that should not be done – either because it involves unacceptable suffering, or because it is of low quality, or both. As shown here, both research quality and suffering have been rated on a four-point scale. 'Quality' is a multidimensional variable which encompasses both scientific excellence and potential practical benefits (see text).

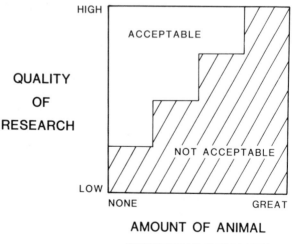

can occasionally raise ethical problems, because the next human they encounter may be carrying a gun rather than binoculars.

Another important point is that for many species, housing individuals in social isolation can cause severe suffering. Social deprivation can be as damaging for some animals as surgical manipulations, even though it leaves no physical scars. Suffering cannot always be judged by overt physical features alone.

Scientists studying animals in the laboratory should have access to professional help on matters of veterinary medicine and should always seek the help of a vet if their animals appear to be ill or suffering.

Everyone studying behaviour has a duty to abide by the spirit (as well as the letter) of any legislation governing animal research, and should strive to minimise the number of animals used in research and the amount of suffering caused to each animal. The number of subjects needed to give a clear result can sometimes be reduced by more careful choice of research design, measurement techniques and statistical analysis (Still, 1982).

F   Further reading

Measurement of bout length is discussed by Slater (1973), Slater & Lester (1982) and Fagen & Young (1978). The issue of ratings versus direct observation in the assessment of personality is discussed by Cairns & Green (1979). The ASAB/ABS Guidelines for the Use of Animals in Research (ASAB/ABS, 1981) are reproduced in full in Appendix 1. For a discussion of animal suffering, see Dawkins (1980) and Wood-Gush (1983, ch. 14). Ethical issues raised by studies of predation and aggression are discussed by Huntingford (1984b). Still (1982) deals with ways in which the numbers of subjects used in animal behaviour experiments can often be reduced without any loss of scientific rigour. Guidelines for the care and management of laboratory animals are found in the *UFAW Handbook* (UFAW, 1976).

Guidelines for the use of animals in research

(From *Animal Behaviour*, **29**, 1–2 (1981))

The use of animals in research raises important ethical issues which are not always made explicit. Studies in laboratory settings necessarily involve keeping animals in cages. Manipulative procedures, and even surgery may be necessary to achieve the aims of the research. Even observation of free-living animals in their natural habitat may involve disruption, particularly if feeding, trapping or marking is involved. While the furthering of scientific knowledge is a proper aim, and may itself advance an awareness of human responsibility towards animal life, the investigator should always weigh any potential gain in knowledge against the adverse consequences for the animals used as subjects, and also for other animals in the ecosystem in the case of field studies. In order to help their members make what are sometimes difficult ethical judgements, the Association for the Study of Animal Behaviour and the Animal Behaviour Society have formed the Ethical and Animal Care committees respectively. These committees have jointly produced the following guidelines for the use of all those who are planning and conducting studies of animal behaviour.

1 The investigator must abide by the spirit as well as the letter of relevant legislation. For those resident in Great Britain, references to laws designed to protect animals are given in the Universities' Federation for Animal Welfare Handbook (UFAW, 1976). In the

USA, both Federal and State legislation may apply: guidance can be obtained from the Code of Federal Regulations (CFR), Title 9, and from the Department of Health Education and Welfare Publication No. (NIH) 78-23 (1978). Workers elsewhere should acquaint themselves with the local requirements.

2   If the animals are confined, constrained, harmed or stressed in any way, the investigator must consider whether the knowledge that may be gained justifies the procedures. Some knowledge is trivial, and experiments must not be done simply because it is possible to do them. Investigators are encouraged to discuss with colleagues both the scientific value of their research proposals and also possible ethical objections. Colleagues who are in a different discipline are especially likely to be helpful since they will not share all the investigator's assumptions.

3   Wherever research involves confining animals, or the use of procedures that are likely to cause pain or discomfort, the research worker should bear in mind that members of some species may be less likely to suffer than others. Choosing an appropriate subject usually requires knowledge of the species' natural history as well as its complexity. Also, knowledge of the animal's previous experience, such as whether or not it has spent a life-time in captivity, can be of profound importance. Alternatives to animal experiments should be considered (Smyth, 1978). Laboratory studies should involve the smallest number of animals necessary to accomplish the research goals properly. The number cannot be precisely specified, but careful thought given to the design of the experiment is important in this context. For instance, if the effects of two factors are to be investigated, the total number of animals required can be greatly reduced if the two factors can be studied concurrently rather than consecutively.

4   Members of endangered species should not be collected or manipulated in the wild except as part of a serious attempt at conservation. Information on threatened species can be obtained from the International Union for the Conservation of Nature, Species Conservation Monitoring Unit, 219C Huntingdon Road, Cambridge CB3 0DL, England. In the USA, rules and regulations pertaining to the Endangered Species Act of 1973 may be found in CFR, Title

50. Lists of endangered species can be obtained by writing to the Office for Endangered Species, US Department of Interior, Fish and Wildlife Service, Washington, DC 20240.

5   Animals should be obtained only from reliable sources. For workers in the UK, advice may be obtained from the MRC's Laboratory Animals Centre, Carshalton, Surrey. In the USA, information on licensed animal dealers can be obtained from the local offices of the US Department of Agriculture (USDA). So far as possible, the investigator should ensure that those responsible for handling the animals en route to the research facilities provide adequate food, water, ventilation, and space, and do not impose undue stress. If animals are trapped in the wild, this should be done in as painless and humane a manner as possible.

6   The experimenter's responsibilities extend also to the conditions under which the animals are kept when not in use. Caging conditions and husbandry practices must meet at the very least minimal recommended requirements. Guidance can be obtained from the UFAW Handbook (1976), and from DHEW Publication No. (NIH) 78-23 (1978).

7   Field workers should strive to disturb their populations as little as possible. In all cases, the investigator should consider, and attempt to minimise, any potentially harmful effects of the study on the population and on other plant and animal species in the community.

These guidelines are intended to supplement the legal requirements in the country and/or State in which the work is carried out.

References

Code of Federal Regulations (1979). *Animals and Animal Products* (Title 9), Subchapter A – Animal Welfare, Parts 1,2,3. Washington, DC: US Government Printing Office.

Code of Federal Regulations (1973). *Wildlife and Fisheries* (Title 50), Chapter 1 (Bureau of Sport Fisheries and Wildlife Service, Fish and Wildlife Sevice, Department of Interior). Washington, DC: US Government Printing Office.

Department of Health, Education and Welfare (1978). *Guide for the Care and Use of Laboratory Animals* (Guidelines prepared by the

Institute of Laboratory Animal Resources). DHEW Publication No. (NIH) 78-23. Washington, DC: US Government Printing Office.

Smyth, D.H. (1978). *Alternatives to Animal Experiments*. London: Scolar Press, Research Defence Society.

Universities Federation for Animal Welfare (1976). *The UFAW Handbook on the Care and Management of Laboratory Animals*, 5th edition. Edinburgh: Churchill Livingstone.

A miniature electronic beeper for time sampling

The beeper was designed by Ray Symonds, of the Cambridge University Department of Zoology, who kindly gave us permission to reproduce the design here.

Circuit description. IC1 is an oscillator whose frequency is determined by a resistor (variable R1 + 10k) and a capacitor (C1 or C2, as selected by switch S2). The output of this oscillator is connected to IC2, which divides the frequency by 65 536. The counter output triggers a multivibrator (IC3) which determines the length of the beep (set by R2 and C3), and the output turns on the transistor TR1. This feeds about 8 volts to IC4, which is the oscillator that produces the audible tone (frequency determined by R3 and C4).

Fig. a2.1. Circuit diagram of a beeper used to denote successive sample points in time sampling. (Capacitor values are in μF.)

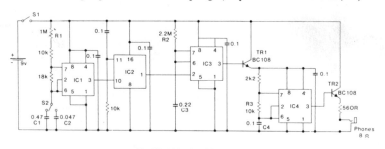

IC1, IC3, IC4 : C-MOS 555

IC2 : C-MOS 4040

This tone is amplified by the transistor TR2 and fed to the earphone. The resistor and capacitor connected to pin 11 of IC2 re-set the counter when the power is first switched on. The beeper has a range of intervals from 6 to 120 s using the 0.047 μF capacitor and 1 to 20 min using the 0.47 μF capacitor, selected by switch S2.

Fig. a2.2. Physical appearance of a beeper.

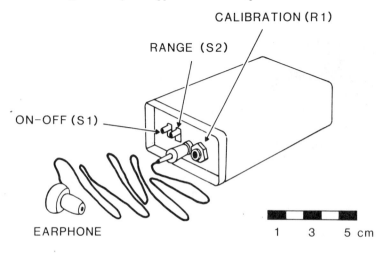

SI units of measurement

The SI system of units (Système International d'Unités) should be used for measurements. The SI system is completely coherent, which means that all derived units are formed by simple multiplication or division of base units without the need for any numerical factors or powers of ten. This distinguishes the SI system from earlier metric systems such as the centimetre-gramme-second (CGS) system, which it superseded. The SI system comprises nine base units, each of which is independently defined, and various other units which are derived by combining two or more base units. The base units, together with some of the more common derived units, are listed in Table a3.1. Some common non-SI units and their SI equivalents are shown in Table a3.2.

Conventions. Each unit is represented by a standard *unit symbol* (e.g., m, s, A, kg), which may be multiplied or divided by other unit symbols or numbers (e.g., 3 m, kg m, m s^{-2}). Unit symbols are algebraic symbols and follow the conventions of algebra. They are not abbreviations, and should never be followed by a full stop or an 's' (to denote plural). The names of units (e.g., metre, second, ampere) are all spelt with a lower case initial letter. Symbols for units named after a person start with an upper case letter (e.g., A for ampere, K for kelvin, Pa for pascal). When a measurement is used as an adjective, the number and unit should be joined by a hyphen (e.g., three-metre tube, 5-m distance, 9-s delay, 6-m^2 area, seventy-kilogram adult, 2-A current).

Standard prefixes (Table a3.3) are used to denote units multiplied by various powers of 10 (e.g., cm = 0.01 m, mA = 0.001 A, km = 1000 m). The prefix and unit symbol together should be treated as a single algebraic entity (e.g., 10 cm s⁻¹ = (10 × 0.01 m) s⁻¹).

Table a3.1. The nine SI base units and some common derived units (BU denotes base unit)

Unit	Symbol	Quantity	Equivalent in base units
metre	m	length	BU
kilogram	kg	mass	BU
second	s	time	BU
ampere	A	electric current	BU
kelvin	K	temperature[a]	BU
mole	mol	amount of substance	BU
candela	cd	luminous intensity	BU
radian	rad	angle	BU
steradian	sr	solid angle	BU
hertz	Hz	frequency[b]	s^{-1}
newton	N	force	$kg\ m\ s^{-2}$
joule	J	energy	N m
watt	W	power	$J\ s^{-1}$
pascal	Pa	pressure	$N\ m^{-2}$
coulomb	C	electric charge	A s
volt	V	potential difference	$J\ C^{-1}$
ohm	Ω	resistance	$V\ A^{-1}$
farad	F	capacitance	$C\ V^{-1}$

[a] In practice, temperature is usually measured in degrees Celsius (°C). One degree Celsius is exactly equal in magnitude to one kelvin, and 0 °C is approximately equal to 273 K. Use of the term centigrade in relation to temperature is incorrect, since a centigrade is a unit of angle. Note that the symbol for the unit of temperature (kelvin) is K, and not °K.
[b] Some authorities suggest that Hz should only be used to describe the frequency of periodic (regularly occurring) phenomena. In the behavioural sciences, Hz should be restricted to the pitch (note) of sounds; for example, vocalisations. When describing the rate of occurrence (frequency) of a behaviour pattern, use the symbol s⁻¹ to avoid confusion.

Table a3.2.   Some commonly used non-SI units

The following units are frequently encountered but, with the exception of °C, their use is not encouraged. Units marked with a dagger (†) are now defined in terms of SI units and the SI equivalent shown is an exact equivalent.

Unit	Symbol	Quantity	SI equivalent
ångström†	Å	length	1.000×10^{-10} m
atmosphere	atm	pressure	1.013×10^5 N m^{-2}
bar†	bar	pressure	1.000×10^5 N m^{-2}
calorie (thermochemical)†	cal	energy	4.184 J
Calorie (nutritional)†	Cal	energy	4.184 kJ
decibel[a]	dB	sound intensity	—
degree Celsius†	°C	temperature	1.000 K
degree Fahrenheit	°F	temperature	0.556 K
hectare†	ha	area	1.000×10^4 m^2
horsepower	hp	power	7.457×10^2 W
inch†	in	length	2.54×10^{-2} m
kilogram-force	kgf	force	9.807 N
litre[b]	l	volume	1.000×10^{-3} m^3
millimetre of mercury	mmHg	pressure	1.333×10^2 N m^{-2}

[a] The decibel (dB) is not strictly a *unit* of sound intensity, but denotes the ratio $10 \log_{10}(I/I')$; where I is the sound intensity (measured in W m^{-2}) and I' is a standard sound intensity (usually 10^{-2} W m^{-2}).
[b] The litre is actually defined in terms of a mass of water, and is not *exactly* equivalent to 1×10^{-3} m^3. To avoid confusion, most authorities recommend using dm^3 instead.

Table a3.3.   Prefixes for SI units

T	tera	10^{12}		h	hecto	10^2		m	milli	10^{-3}
G	giga	10^9		da	deka	10		μ	micro	10^{-6}
M	mega	10^6		d	deci	10^{-1}		n	nano	10^{-9}
k	kilo	10^3		c	centi	10^{-2}		p	pico	10^{-12}

REFERENCES

In addition to the references cited in the text or in the Further
Reading sections, we have included some references (marked †)
on a variety of methodological issues. References marked * are
particularly recommended.

Abbey, H. & Howard, E. (1973). Statistical procedure in
developmental studies on species with multiple offspring.
Developmental Psychobiology, **6**, 329–335.
A reminder that measurements of littermates are likely to be correlated
and so cannot be assumed to be statistically independent. Use litter-
mean values as data points.

Aldrich-Blake, F.P.G. (1970). Problems of social structure in
forest monkeys. In *Social Behaviour in Birds and Mammals.
Essays on the Social Ethology of Animals and Man* (ed. by
J.H. Crook), pp. 79–101. London & New York: Academic
Press.
Considers various problems of bias in field studies, such as poor visibility
leading to difficulties in seeing and following subjects, and differences
in the visibility of subjects in different habitats. Emphasises that short
studies, especially those of species living in conditions where visibility
is poor, can be misleading. Poor visibility reduces the quantity as well
as the quality of behavioural data.

Altman, P.L. & Dittmer, D.S. (eds.) (1972–1974). *Biology Data
Book*, 2nd edition (3 volumes). Bethesda, Maryland: Federation
of American Societies for Experimental Biology.
A comprehensive source of information on various aspects of the biology
of different species. Includes sections on reproduction, growth and
development (Vol. I); tolerance to temperature and other environmental
factors, sensory and neural biology (Vol. II); nutrition, metabolism,
blood and other body fluids (Vol. III).

*Altmann, J. (1974). Observational study of behavior: sampling methods. *Behaviour*, **49**, 227–267.
An important paper, discussing observational methods for studying groups of animals, mainly applicable to field studies; arguably wrong in its outright condemnation of one-zero sampling; primarily concerned with field research methods.

Altmann, J. (1980). *Baboon Mothers and Infants*. Cambridge, Massachusetts: Harvard University Press.
A good example of rigorous field study methods. See ch. 3 on methods and appendix 4 for examples of how observational categories of behaviour are defined.

†Altmann, S.A. & Altmann, J. (1977). On the analysis of rates of behaviour. *Animal Behaviour*, **25**, 364–372.
Gives formulae for calculating expected frequencies of behaviour for particular age, sex and other classes, in groups where the proportions of individuals in each class are unequal and vary with time.

Amlaner, C.J. Jr (1978). Biotelemetry from free-ranging animals. In *Animal Marking. Recognition Marking of Animals in Research* (ed. by B. Stonehouse), pp. 205–228. London: Macmillan.
Deals with the measurement of physiological variables, such as body temperature, respiration, ECG, EEG, EMG, heart rate, blood flow and heat flow in free-ranging animals using radio telemetry.

Amlaner, C.J. (1981). Techniques of study. In *The Oxford Companion to Animal Behaviour* (ed. by D. McFarland), pp. 544–550. Oxford: Oxford University Press.
Reviews a variety of methods, including capture and identification of animals, recording techniques and telemetry. See also entries in the *Oxford Companion* under Classification, Field Studies and Laboratory Studies.

Amlaner, C.J. & Macdonald, D.W. (eds.) (1980). *A Handbook on Biotelemetry and Radio Tracking*. Oxford: Pergamon Press.
A comprehensive, multi-authored volume dealing with all aspects of telemetry and radio-tracking, including design, performance and construction of apparatus, data acquisition and analysis. Gives many examples of applications of biotelemetry and radio tracking in studies of mammals, birds, fish and invertebrates.

†Anderson, D.J. (1982). The home range: a new nonparametric estimation technique. *Ecology*, **63**, 103–112.
Presents a new, nonparametric technique for estimating home range size from a series of separate observations of position.

Andersson, M. (1974). Temporal graphical analysis of behaviour sequences. *Behaviour*, **51**, 38–48.
Presents a graphical method for summarising behavioural sequences

(such as displays and other social behaviour) and analysing temporal associations between the different behaviour patterns in a sequence.

Appleby, M.C. (1983). The probability of linearity in hierarchies. *Animal Behaviour*, **31**, 600–608.
Deals with methodological problems concerning dominance hierarchies, showing that even when no dominance relations exist, apparent linear hierarchies will often arise by chance.

*ASAB/ABS (1981). Guidelines for the use of animals in research. *Animal Behaviour*, **29**, 1–2.
Reprinted here as Appendix 1. A new, extended set of guidelines is published in *Animal Behaviour*, **34**, 315–318 (1986).

Baerends, G. (1981). Field studies. In *The Oxford Companion to Animal Behaviour* (ed. by D. McFarland), pp. 183–189. Oxford: Oxford University Press.
A concise account of why field studies are necessary.

Bailey, N.T.J. (1981). *Statistical Methods in Biology*, 2nd edition. London: Hodder & Stoughton.
A short, introductory text on statistics.

Barlow, G.W. (1977). Modal action patterns. In *How Animals Communicate* (ed. by T.A. Sebeok), pp. 98–134. Bloomington: Indiana University Press.
Makes the point that 'fixed' action patterns are actually variable in form, making the term 'modal action pattern' more appropriate.

Bateson, P.P.G. (1964). Changes in chicks' responses to novel moving objects over the sensitive period for imprinting. *Animal Behaviour*, **12**, 479–489.

†Bateson, P.P.G. (1968). Ethological methods of observing behavior. In *Analysis of Behavioral Change* (ed. by L. Weiskrantz), pp. 389–399. New York: Harper & Row.
A general review of observational methods.

Bateson, P.P.G. (1977). Testing an observer's ability to identify individual animals. *Animal Behaviour*, **25**, 247–248.
Describes how the ability of an experienced observer (D.K. Scott) to recognise some 450 individual Bewick's swans, using variations in their bill patterns, was tested. In one test, Scott successfully identified from photographs 29 out of 30 individual swans, taking a median time of 2.0 s to identify each bird (see also Scott, 1978).

Bateson, P. [P.G.] (1979*a*). How do sensitive periods arise and what are they for? *Animal Behaviour*, **27**, 470–486.
Discusses developmental and functional aspects of sensitive periods: phases during ontogeny when a factor is especially likely to exert an effect on the developing organism.

Bateson, P. [P.G.] (1979*b*). Brief exposure to a novel stimulus during imprinting in chicks and its influence on subsequent preferences. *Animal Learning and Behavior*, **7**, 259–262.

An example of how repeated exposure to choice stimuli can affect the very preferences the experiment is attempting to measure.

Batschelet, E. (1979). *Introduction to Mathematics for Life Scientists*, 3rd edition. New York: Springer-Verlag.
A widely used text on mathematics for biologists.

Beach, F.A. (1950). The snark was a boojum. *American Psychologist,* **5**, 115–124. [Reprinted in McGill, 1977.]
A classic paper, drawing attention to the narrow theoretical and empirical basis of 'comparative' psychology in the 1940s. Beach, who described American psychology as the study of learning in the white rat and college sophomore, made an influential plea for psychologists to diversify the problems and species they studied. Since Beach's article was written, comparative psychology has developed beyond the narrow view that learning is the only important problem in the study of animal behaviour and that the white rat (or indeed any single species) is a 'representative' organism.

†Bekoff, M. (1976). The ethics of experimentation with non-human subjects: should Man judge by vision alone? *The Biologist*, **58**, 30–31.
Makes the point that experimental manipulations such as social deprivation, which do not leave obvious physical signs in the way that surgical manipulations do, can in many ways be just as damaging. 'What you see is not necessarily what the animal gets.'

Bekoff, M. (1977). Quantitative studies of three areas of classical ethology: social dominance, behavioral taxonomy, and behavioral variability. In *Quantitative Methods in the Study of Animal Behavior* (ed. by B.A. Hazlett), pp. 1–46. New York: Academic Press.
A useful discussion of quantitative methods in the study of dominance hierarchies, classification of behaviour, 'fixed action patterns' and sequence analysis.

†Bekoff, M. & Mech, L.D. (1984). Simulation analyses of space use: home range estimates, variability, and sample size. *Behavior Research Methods, Instrumentation, and Computers*, **16**, 32–37.
Considers how sample size may influence the measurement of home range and territory size when using radio tracking.

Berger, J. (1979). Weaning conflict in desert and mountain bighorn sheep (*Ovis canadensis*): an ecological interpretation. *Zeitschrift für Tierpsychologie*, **50**, 188–200.
An example of a field study of intraspecific variation in behaviour. Suckling behaviour and weaning were studied in three natural populations of sheep and the differences in behaviour related to their different habitats.

Bernstein, I.S. (1981). Dominance: the baby and the bathwater. *The Behavioral and Brain Sciences*, **4**, 419–457.
A general review of the concept of dominance, with comments from other scientists.

Block, J. (1977). Advancing the psychology of personality: paradigmatic shift or improving the quality of research? In *Personality at the Crossroads. Current Issues in Interactional Psychology* (ed. by D. Magnusson & N.S. Endler), pp. 37–63. Hillsdale, NJ: Lawrence Erlbaum.
Reviews the issue of personality assessment in humans.

†Blurton Jones, N. (ed.) (1972). *Ethological Studies of Child Behaviour*. Cambridge: Cambridge University Press.
An example of how ethological methods of observation have been applied to the study of children's behaviour.

Borgerhoff Mulder, M. & Caro, T.M. (1985). The use of quantitative observational techniques in anthropology. *Current Anthropology*, **26**, 323–335.
Outlines the applications of direct observation of behaviour in anthropology. Covers many basic issues, such as sources of bias, observer reliability and the problem of definition.

Bowlby, J. (1951). *Maternal Care and Mental Health*. Geneva: World Health Organization.

Box, G.E.P. & Jenkins, G.M. (1970). *Time Series Analysis. Forecasting and Control*. San Francisco: Holden-Day.
Reviews methods for analysing inter-dependent observations that occur in the form of a series over time; for example, spectral analysis and autocorrelation.

Boyd, R. & Silk, J.B. (1983). A method for assigning cardinal dominance ranks. *Animal Behaviour*, **31**, 45–58.
Dominance hierarchies are normally described in terms of each individual's rank; i.e., dominance is measured on an ordinal scale. This paper presents a method for assigning cardinal ranks; i.e., measuring dominance status on an interval scale. This allows the difference in rank of two individuals to be quantified and tested for statistical significance, and enables parametric statistics to be used for analysing dominance scores.

Broom, D.M. (1979). Methods of detecting and analysing activity rhythms. *Biology of Behaviour*, **4**, 3–18.
Reviews the issue of ultradian, circadian and other rhythms in behaviour and describes methods for detecting the presence of such periodicities in behavioural data.

Broom, D.M. (1980). Activity rhythms and position preferences of domestic chicks which can see a moving object. *Animal Behaviour*, **28**, 201–211.

An example of the analysis of behavioural rhythms. The activity levels of isolated domestic chicks were recorded every 30 min for several days after hatching. Data were analysed using autocorrelation, spectral analysis and multiple regression analysis. Most chicks were found to show ultradian rhythmicities (with periods of 1.5–4 h and 30 min) superimposed over a circadian rhythm.

Broom, D.M. (1981). *Biology of Behaviour. Mechanisms, Functions and Applications.* Cambridge: Cambridge University Press.

A good general text on animal behaviour. Includes an extensive account of the practical applications of ethology.

Burley, N., Krantzberg, G. & Radman, P. (1982). Influence of colour-banding on the conspecific preferences of zebra finches. *Animal Behaviour*, **30**, 444–455.

Coloured plastic leg bands affected the sexual attractiveness of zebra finches, implying that marking birds for identification purposes can have substantial effects on their behaviour and that of conspecifics.

Cairns, R.B. (ed.) (1979). *The Analysis of Social Interactions. Methods, Issues, and Illustrations.* Hillsdale, NJ: Lawrence Erlbaum.

A multi-authored volume, dealing with the theoretical and practical issues of measuring social interactions, mainly in humans. Also includes discussions of the analysis of sequences (ch. 4) and the assessment of personality using observer ratings (appendix A).

Cairns, R.B. & Green, J.A. (1979). How to assess personality and social patterns: observations or ratings? In *The Analysis of Social Interactions. Methods, Issues, and Illustrations* (ed. by R.B. Cairns), pp. 209–226. Hillsdale, NJ: Lawrence Erlbaum.

Discusses the relative merits and drawbacks of the two main methods of assessing personality and social interactions: ratings and direct observation. Concludes that both are useful, but for different purposes.

Campbell, D.J. & Shipp, E. (1974). Spectral analysis of cyclic behaviour with examples from the field cricket *Teleogryllus commodus* (Walk.). *Animal Behaviour*, **22**, 862–875.

Exemplifies the use of spectral analysis to detect cyclicities in behaviour. Data describing the migratory and locomotor behaviour of crickets exhibited complex rhythms with multiple components.

Cane, V. (1961). Some ways of describing behaviour. In *Current Problems in Animal Behaviour* (ed. by W.H. Thorpe & O.L. Zangwill), pp. 361–388. Cambridge: Cambridge University Press.

Deals with the independence of categories.

Cane, V. R. (1978). On fitting low-order Markov chains to behaviour sequences. *Animal Behaviour*, **26**, 332–338.
Shows how the description of behaviour sequences can be simplified by grouping together related categories of behaviour using statistical techniques. Discusses Markov-chain analysis, a principal method of sequence analysis.

Caro, T.M., Roper, R., Young, M. & Dank, G.R. (1979). Inter-observer reliability. *Behaviour*, **69**, 303–315.
Explains how inter-observer reliability can be measured using a correlation or index of concordance and discusses various factors that affect observer reliability, including problems of definition. Examples are drawn from observational studies of kittens and children.

†Charlesworth, W.R. (1978). Ethology: its relevance for observational studies of human adaptation. In *Observing Behavior*, Vol I, *Theory and Applications in Mental Retardation* (ed. by G.P. Sackett), pp. 7–32. Baltimore: University Park Press.
Discusses the ethological approach and its contributions, both conceptual and methodological, to the study of human behaviour. Emphasises the importance of longitudinal observation in studies of development.

Chase, I.D. (1974). Models of hierarchy formation in animal societies. *Behavioral Science*, **19**, 374–382.
Deals with methods of describing dominance hierarchies, including Landau's index of linearity (h).

Chatfield, C. & Lemon, R.E. (1970). Analysing sequences of behavioural events. *Journal of Theoretical Biology*, **29**, 427–445.
Reviews methods for analysing sequential dependencies in behavioural time series and considers the relations between the chi square goodness-of-fit and information theory.

Cheney, D.L. & Seyfarth, R.M. (1982). How vervet monkeys perceive their grunts: field playback experiments. *Animal Behaviour*, **30**, 739–751.
A good example of how experiments can be carried out in the field. Field playback experiments showed that grunt vocalisations produced by free-ranging vervet monkeys in social situations are perceived as at least four different vocalisations by other vervet monkeys, even though humans cannot differentiate between them. Each grunt seems to have a specific meaning, which is conveyed more by its acoustic properties than by the context in which it occurs.

Chow, I.A. & Rosenblum, L.A. (1977). A statistical investigation of the time-sampling methods in studying primate behavior. *Primates*, **18**, 555–563.

Provides techniques for obtaining the best estimates of frequencies and durations from instantaneous or one-zero time sampling records.

Clarke, G.M. (1980). *Statistics and Experimental Design*, 2nd edition. London: Edward Arnold.
A succinct account for biologists of basic issues of statistics and experimental design.

Clutton-Brock, T.H., Guinness, F.E. & Albon, S.D. (1982). *Red Deer. Behavior and Ecology of Two Sexes*. Chicago: University of Chicago Press/Edinburgh: Edinburgh University Press.
A detailed and long-term field study of a large mammal; provides an excellent example of how the behaviour of free-living animals can be measured.

Cochran, W.G. (1977). *Sampling Techniques*, 3rd edition. New York: Wiley.
Explains fully the theory and practice of sampling and surveying, including simple random sampling (ch. 2), estimation of sample size (ch. 4) and stratified random sampling (ch. 5).

Cohen, J. (1977). *Statistical Power Analysis for the Behavioral Sciences*, revised edition. New York: Academic Press.
Deals with methods for estimating the required sample size under a variety of circumstances.

Colgan, P.W. (ed.) (1978). *Quantitative Ethology*. New York: Wiley.
An advanced, multi-authored text covering several quantitative methods in detail; useful for reference, particularly on multivariate statistical methods. See chapters by Slater (data collection), Fagen & Young (temporal patterning), DeGhett (cluster analysis) and Frey & Pimentel (principal component analysis and factor analysis).

Conover, W.J. (1980). *Practical Nonparametric Statistics*, 2nd edition. New York: Wiley.
Covers most nonparametric tests, including many developed since Siegel's book was written.

Cramp, S. & Simmons, K.E.L. (eds.) (1977–1985). *The Birds of the Western Palearctic. Handbook of the Birds of Europe the Middle East and North Africa*. Vol. I (1977); Vol. II (1980); Vol. III (1983); Vol. IV (1985). Oxford: Royal Society for the Protection of Birds/Oxford University Press.
A comprehensive and beautifully presented source of information on bird species.

Crook, J.H. (1970). The socio-ecology of primates. In *Social Behaviour In Birds and Mammals. Essays on the Social Ethology of Animals and Man* (ed. by J.H. Crook), pp. 103–166. London & New York: Academic Press.

A wide-ranging review of field studies of primate social behaviour. Makes the distinction between 'group' (social unit of known composition) and 'party' (social unit of uncertain composition).

Cruze, W.W. (1935). Maturation and learning in chicks. *Journal of Comparative Psychology*, **19**, 371–408.
A classic experiment, showing that developmental changes in behaviour are due both to internal changes and experience.

Cullen, J.M. (1972). Some principles of animal communication. In *Non-Verbal Communication* (ed. by R.A. Hinde), pp. 101–122. Cambridge: Cambridge University Press.
The two church clocks analogy, illustrating the principle that correlation does not imply causation.

Davies, N.B. (1982). Behaviour and competition for scarce resources. In *Current Problems in Sociobiology* (ed. by King's College Sociobiology Group), pp. 363–380. Cambridge: Cambridge University Press.
A clear account of how individual differences in behaviour may have evolved. Shows how the selfish gene approach, the use of game theory and detailed studies of known individuals in wild populations of animals have emphasised that individuals within a species behave in markedly different ways when competing for scarce resources such as food or a mate. Three types of alternative strategy are outlined: those conditional on some feature of the environment; those conditional on the individual's phenotype; and frequency-dependent equilibria.

Dawkins, M.S. (1980). *Animal Suffering. The Science of Animal Welfare*. London: Chapman & Hall.
A thoughtful scientific analysis of animal welfare, showing that it can be studied empirically. Evaluates various empirical methods for assessing suffering when there are no physical signs of ill-health or injury.

Dawkins, M.S. (1983). The organisation of motor patterns. In *Animal Behaviour*, Vol. 1, *Causes and Effects* (ed. by T.R. Halliday & P.J.B. Slater), pp. 75–99. Oxford: Blackwell Scientific Publications.
Reviews units of behaviour, fixed action patterns and sequences of behaviour.

Dawkins, R. (1971). A cheap method of recording behavioural events, for direct computer-access. *Behaviour*, **40**, 162–173.
Describes a cheap and simple electronic 'organ' for recording behaviour. The keyboard generates a pulse of different frequency for each category. Information is stored on an ordinary tape-recorder prior to decoding and analysis by a computer. More than 60 keys can be used, with a time resolution of 0.1 s, although the device does not allow two events to be recorded simultaneously.

DeGhett, V.J. (1978). Hierarchical cluster analysis. In *Quantitative Ethology* (ed. by P.W. Colgan), pp. 115–144. New York: Wiley.
An introduction to hierarchical cluster analysis for ethologists. Cluster analysis is a general term for multivariate statistical techniques that seek groupings in data.

†Dixon, K.R. & Chapman, J.A. (1980). Harmonic mean measure of animal activity areas. *Ecology*, **61**, 1040–1044.
Presents a method of calculating areas of activity (home range, territory size, foraging area, etc.) and centres of activity, based on the harmonic mean of an areal distribution. Reviews other techniques of measuring home ranges, territory sizes and activity centres.

Dlhopolsky, J.G. (1983). Limitations of high-level microcomputer languages in software designed for psychological experimentation. *Behavior Research Methods & Instrumentation*, **15**, 459–464.
Timing to millisecond precision cannot be done with high-level languages such as BASIC or Pascal and requires machine language software. Fortunately, it is seldom necessary for recording behavioural observations.

Douglas, J.M. & Tweed, R.L. (1979). Analysing the patterning of a sequence of discrete behavioural events. *Animal Behaviour*, **27**, 1236–1252.
Reviews conventional techniques for analysing sequences of behaviour (such as cluster analysis, mutual replaceability, 'melody' detection and Markov analysis) and presents some newer, simpler techniques.

Douglas, J.W.B. (1975). Early hospital admissions and later disturbances of behaviour and learning. *Developmental Medicine and Child Neurology*, **17**, 456–480.

Drickamer, L.C. & Vessey, S.H. (1982). *Animal Behavior. Concepts, Processes, and Methods.* Boston: Willard Grant Press.
A general text on animal behaviour; unusual among modern books in that it includes discussions of methodological issues. See especially ch. 3 ('Approaches and methods').

†Drummond, H. (1981). The nature and description of behavior patterns. In *Perspectives in Ethology*, Vol. 4 (ed. by P.P.G. Bateson & P.H. Klopfer), pp. 1–33. New York: Plenum Press.
Lengthy discussion of some of the issues raised by splitting a continuous stream of behaviour up into discrete, named units; for example, the problem of 'natural' units of behaviour.

Dunbar, R.I.M. (1976). Some aspects of research design and their implications in the observational study of behaviour. *Behaviour*, **58**, 78–98.

Discusses various ways in which behaviour can be sampled and described, dealing mainly with field studies of primate social behaviour. Shows, for example, that frequency and duration measures can give different pictures of the same behaviour.

Edwards, A.W.F. (1972). *Likelihood.* Cambridge: Cambridge University Press. (Re-issued as a Cambridge Science Classic, 1984.)
Argues that the appropriate basis for statistical inference is not the traditional one of probability, but that of likelihood: a measure of relative support among different hypotheses.

Eisenberg, J.F. (1981). *The Mammalian Radiations. An Analysis of Trends in Evolution, Adaptation, and Behaviour.* London: Athlone Press.
A comprehensive and synthetic review of the morphology, ecology and behaviour of mammals, giving much information on species diversity. Section 4 deals specifically with behaviour and includes a description and classification of the behaviour patterns common to many terrestrial mammals. The appendices include extensive compilations of data on the reproductive rate, litter size, inter-birth interval, longevity, brain size and metabolic rate of many species. The book also has a large bibliography on mammalian natural history.

Eltringham, S.K. (1978). Methods of capturing wild animals for marking purposes. In *Animal Marking. Recognition Marking of Animals in Research* (ed. by B. Stonehouse), pp. 13–23. London: Macmillan.
Covers techniques for capturing wild birds and mammals, including traps, nets and stupefying drugs.

†Everitt, B.S. (1977). *The Analysis of Contingency Tables.* London: Chapman & Hall.

†Fagen, R.M. (1978). Repertoire analysis. In *Quantitative Ethology* (ed. by P.W. Colgan), pp. 25–42. New York: Wiley.
Describes methods, such as regression analysis, used for estimating the total number of categories needed to describe a species' entire behavioural repertoire, including rare behaviour patterns.

Fagen, R.M. & Mankovich, N.J. (1980). Two-act transitions, partitioned contingency tables, and the 'significant cells' problem. *Animal Behaviour*, **28**, 1017–1023.
Discusses the vexed problem of two-way contingency tables in the analysis of behavioural data, such as sequences and dominance relationships.

Fagen, R.M. & Young, D.Y. (1978). Temporal patterns of behaviors: durations, intervals, latencies, and sequences. In *Quantitative Ethology* (ed. by P.W. Colgan), pp. 79–114. New York: Wiley.

Reviews methods for analysing temporal patterns and sequences, such as log survivorship and Markov chain analysis.

†Fassnacht, G. (1982). *Theory and Practice of Observing Behaviour*. London: Academic Press.
Aimed primarily at social psychologists and sociologists and written with more emphasis on the theory rather than the practice of recording behaviour. Includes discussions of the nature of objectivity, the relations between observation and experimentation and recognising units of behaviour. Ethological methods are considered in ch. 6.

Feaver, J., Mendl, M. & Bateson, P. (1986). A method for rating the individual distinctiveness of domestic cats. *Animal Behaviour*, **34**, (in press).
Two observers watched 14 adult female domestic cats for three months and independently rated them on 18 aspects of each cat's behavioural style. Inter-observer reliability correlations were statistically significant for 15 of the 18 items, and acceptably large ($r > 0.7$) for 7 items. Ratings were compared with direct observations where possible; for 5 out of 6 items, the rating and direct measurement were significantly correlated.

Flowers, J.H. (1982). Some simple Apple II software for the collection and analysis of observational data. *Behavior Research Methods & Instrumentation*, **14**, 241–249.
Listings of some very simple BASIC programs for event-recording using a standard laboratory microcomputer.

Flowers, J.H. & Leger, D.W. (1982). Personal computers and behavioral observation: an introduction. *Behavior Research Methods & Instrumentation*, **14**, 227–230.

Folks, J.L. (1981). *Ideas of Statistics*. New York: Wiley.
A concise introduction to the history and major concepts of statistics and experimental design.

Francis, S.H. (1966). An ethological study of mentally retarded individuals and normal infants. Unpublished PhD dissertation, University of Cambridge.

Frey, D.F. & Pimentel, R.A. (1978). Principal component analysis and factor analysis. In *Quantitative Ethology* (ed. by P.W. Colgan), pp. 219–245. New York: Wiley.
Discusses applications of these major types of multivariate statistical analysis to behavioural data.

Gans, C. (1978). All animals are interesting! *American Zoologist*, **18**, 3–9.
Contrasts the 'principles' approach to biology (picking a species to study because it suits a particular problem) with studying a species or group of species for their own sake ('scientific natural history'). Argues persuasively that we often don't know enough about different

species to make the right choices, and defends the natural history approach.

Ghiselli, E.E., Campbell, J.P. & Zedeck, S. (1981). *Measurement Theory for the Behavioral Sciences.* San Francisco: W.H. Freeman.
Explains the major issues of measurement in psychology. Includes chapters on the concept of correlation (ch. 5), combining component variables into a composite (ch. 7) and the reliability and validity of measurement (chs. 8, 9, 10).

†Golani, I. (1976). Homeostatic motor processes in mammalian interactions: a choreography of display. In *Perspectives in Ethology*, Vol. 2 (ed. by P.P.G. Bateson & P.H. Klopfer), pp. 69–134. New York: Plenum Press.
Describes the complicated Eshkol-Wachmann dance notation for recording sequences of movement in fine detail.

Gottman, J.M. (1978). Nonsequential data analysis techniques in observational research. In *Observing Behavior*, Vol II, *Data Collection and Analysis Methods* (ed. by G.P. Sackett), pp. 45–61. Baltimore: University Park Press.
Describes how statistical relationships between different categories of behaviour can be discovered using multivariate techniques such as principal component analysis or factor analysis.

Gottman, J.M. (1981). *Time-Series Analysis. A Comprehensive Introduction for Social Scientists.* Cambridge: Cambridge University Press.
Covers many aspects of statistical analysis of time-related data.

†Govil, A.K. (1984). *Definitions and Formulae in Statistics*, 2nd edition. London: Macmillan.
A source-book of definitions and formulae, covering most commonly-used statistical methods.

Griffin, D.R. (1984). *Animal Thinking.* Cambridge, Massachusetts: Harvard University Press.
Considers the fascinating issues of consciousness and cognition in non-human species.

Hailman, J.P. (1973). Fieldism. *BioScience*, **23**, 149.
A short, spirited and humorous attack on various forms of discrimination against field workers ('fieldism').

Halliday, T.R. (1976). The libidinous newt. An analysis of variations in the sexual behaviour of the male smooth newt, *Triturus vulgaris. Animal Behaviour*, **24**, 398–414.
An example of how a multivariate statistical technique (principal component analysis) can be used to group together several correlated measures of behaviour to give a single, composite score that explains much of the variation in behaviour.

†Hansell, M.H. & Aitken, J.J. (1977). *Experimental Animal Behaviour*. Glasgow: Blackie.

A summary of the principal methods for recording behaviour, plus a collection of laboratory practical exercises.

Harcourt, A.H. (1978). Activity periods and patterns of social interaction: a neglected problem. *Behaviour*, **66**, 121–135.

Considers how the time of day at which behaviour is observed affects the conclusions drawn. For example, a field study of mountain gorillas found that the nature as well as the overall amount of their social behaviour changed during the course of the day, according to their predominant maintenance activity (resting, moving or feeding).

Harcourt, A.H. & Stewart, K. J. (1984). Gorillas' time feeding: aspects of methodology, body size, competition and diet. *African Journal of Ecology*, **22**, 207–215.

Field observations using focal sampling found that gorillas spend more time feeding than was previously thought. Earlier studies had used scan sampling and, since gorillas tend to move out of sight when feeding, had consequently underestimated the time spent feeding.

†Harlow, H.F., Gluck, J.P. & Suomi, S.J. (1972). Generalization of behavioral data between nonhuman and human animals. *American Psychologist*, **27**, 709–716.

The authors conclude this entertaining essay on the justification for generalizing from nonhuman to human research as follows: '...one cannot generalize, but one must. If the competent do not wish to generalize, the incompetent will fill the field.' Makes the point that the only way to test the limits of interspecies generalisations is empirically.

Harnett, D.L. (1982). *Statistical Methods*, 3rd edition. Reading, Massachusetts: Addison-Wesley.

A good general statistics text.

†Harré, R. & Lamb, R. (eds.) (1983). *The Encyclopedic Dictionary of Psychology*. Cambridge, Massachusetts: MIT Press.

A useful reference book, covering a wide range of topics in psychology and animal behaviour.

†Hazlett, B.A. (ed.) (1977). *Quantitative Methods in the Study of Animal Behavior*. New York: Academic Press.

A multi-authored collection of chapters on advanced topics, including information theory, multivariate analysis and the treatment of non-stationary sequences.

Hinde, R.A. (1970). *Animal Behaviour. A Synthesis of Ethology and Comparative Psychology*, 2nd edition. New York: McGraw-Hill.

Still an excellent general text on animal behaviour, despite being somewhat out of date. See ch. 1 (aims and methods) and ch. 2 (description and classification of behaviour).

*Hinde, R.A. (1973). On the design of check-sheets. *Primates*, **14**, 393–406.
Sound advice on the design and use of check sheets and recording methods in general, including the nature of behavioural units, the number, scope and exclusiveness of categories and the various recording rules.

Hinde, R.A. (1979*a*). Report on replies to a questionnaire concerning the experiences of recent graduates and others abroad. *Supplement to Primate Eye No. 11*.
Summarises the results of a questionnaire sent to field biologists and discusses the various physical and psychological problems faced by those doing field studies abroad.

Hinde, R.A. (1979*b*). *Towards Understanding Relationships*. London: Academic Press.

Hinde, R.A. (1982). *Ethology. Its Nature and Relations with Other Sciences*. New York: Oxford University Press.
A succinct introduction to the central ideas of ethology. The second half discusses the relations between ethology and other behavioural sciences such as behavioural ecology, neurobiology, experimental and comparative psychology, social and developmental psychology, anthropology and psychiatry.

Hinde, R.A. (ed.) (1983). *Primate Social Relationships. An Integrated Approach*. Oxford: Blackwell Scientific Publications.
A collection of short sections on many aspects of primate social behaviour, with useful accounts of methods. See especially the sections by R.A. Hinde on general issues in describing social behaviour (pp. 1–7 & 17–20), P.C. Lee on methods for studying primate behaviour (pp. 8–16), J. Stevenson-Hinde on assessing individual characteristics (pp. 28–35) and R.A. Hinde (pp. 334–339) on the extent to which the concepts and methods used for studying non-human species can be applied to human behaviour.

Hinde, R.A. & Atkinson, S. (1970). Assessing the roles of social partners in maintaining mutual proximity, as exemplified by mother–infant relations in rhesus monkeys. *Animal Behaviour*, **18,** 169–176.
Presents an index for quantifying how much one individual in a dyadic relationship is responsible for maintaining mutual proximity (see 8.E).

Hollenbeck, A.R. (1978). Problems of reliability in observational research. In *Observing Behavior*, Vol. II, *Data Collection and Analysis Methods* (ed. by G.P. Sackett), pp. 79-98. Baltimore: University Park Press.

A comprehensive discussion of the concept of reliability in measuring behaviour and various ways of assessing it.

Holm, R.A. (1978). Techniques of recording observational data. In *Observing Behavior*, Vol. II, *Data Collection and Analysis Methods* (ed. by G.P. Sackett), pp. 99–108. Baltimore: University Park Press.
A concise review of various ways of recording behavioural observations, including verbal descriptions, event recorders and check sheets.

Huntingford, F. (1984a). *The Study of Animal Behaviour*. London: Chapman & Hall.
Ch. 2 (pp. 12–46) covers the description and measurement of behaviour. Incidentally, this is one of the best general texts on animal behaviour.

Huntingford, F.A. (1984b). Some ethical issues raised by studies of predation and aggression. *Animal Behaviour*, **32**, 210–215.
Considers the ethical issues arising when predator-prey or aggressive encounters are artificially staged as part of an experimental study. Argues that, where such encounters are necessary, the use of model predators should be considered; the number of subjects used and the duration of the experiment should be kept to a minimum; care should be taken to ensure that the amount of suffering is minimised while the amount of information obtained is maximised; and the theoretical interest and justification for the experiment should be examined critically.

*Hutt, S.J. & Hutt, C. (1970). *Direct Observation and Measurement of Behavior*. Springfield, Illinois: Thomas.
A thoughtful, concise and clear account of ethological methods for the direct observation and recording of behaviour, illustrated with examples from studies of children, psychiatric patients and animals. Includes references to several pioneering studies of children's behaviour, carried out in the 1920s and 30s, which used observational methods.

Jaynes, J. (1969). The historical origins of 'ethology' and 'comparative psychology'. *Animal Behaviour*, **17**, 601–606.
Traces the historical development of ethology (the biological study of behaviour) and psychology.

Kavaliers, M. (1980). Pineal control of ultradian rhythms and short-term activity in a cyprinid fish, the lake chub, *Couesius plumbeus*. *Behavioral and Neural Biology*, **29**, 224–235.
An example of ultradian rhythmicity in behaviour.

Kieras, D.E. (1981). Effective ways to dispose of unwanted time and money with a laboratory computer. *Behavior Research Methods & Instrumentation*, **13**, 145–148.

Illustrates four major errors that can be made when acquiring computer equipment: avoiding advice, building your own hardware, buying from a fly-by-night manufacturer and being ignorant of the machine.

Kraemer, H.C. (1979*a*). One-zero sampling in the study of primate behaviour. *Primates*, **20**, 237–244.
Shows that one-zero sampling measures a linear combination of total duration (or proportion of time spent performing the behaviour pattern) and frequency (number of bouts per unit time), the relative contributions of the two depending on the length of the sample interval. As the sample interval becomes shorter, one-zero sampling increasingly reflects the total duration (or proportion of time). Also shows that instantaneous sampling gives an unbiased estimate of total duration (or proportion of time), the precision of which improves as the sample interval becomes shorter. Recommends that alternatives to one-zero sampling should be used whenever possible.

Kraemer, H.C. (1979*b*). Ramifications of a population model for κ as a coefficient of reliability. *Psychometrika*, **44**, 461–472.
Proposes a population model for the definition of the kappa coefficient of reliability (κ), analogous to that for the reliability correlation coefficient. Kappa, like a reliability correlation, directly indicates the loss of precision or power of statistical procedures as a result of observation error. Examines the interpretation of kappa as a measure of diagnostic reliability in characterising an individual, and the effect of reliability on bias, precision and statistical power. Also considers factors affecting kappa. Suggests two strategies for overcoming the effects of low reliability: increasing the sample size; and combining multiple, unreliable measures to yield one, reliable measure.

*Kraemer, H.C. (1981). Coping strategies in psychiatric clinical research. *Journal of Consulting and Clinical Psychology*, **49**, 309–319.
Invaluable practical advice from an eminent statistician on measurement, research design and statistical analysis, of relevance to anyone studying behaviour. She suggests the best 'coping strategies' for tackling research within the limitations of the real world rather than the idealised world of the statistics textbook.

Kraemer, H.C., Hole, W.T. & Anders, T.F. (1984). The detection of behavioral state cycles and classification of temporal structure in behavioral states. *Sleep*, **7**, 3–17.
Proposes an objective method for classifying behavioural cycles in terms of their temporal structure.

†Krebs, H.A. (1975). The August Krogh principle: 'For many problems there is an animal on which it can be most conveniently studied.' *Journal of Experimental Zoology*, **194**, 221–226.

The idea stated in the title is illustrated with examples from ethology, biochemistry, physiology, cell biology, experimental medicine and botany. For example, the ethologist Niko Tinbergen found the three-spine stickleback particularly suitable for study because it maintains its normal behaviour in captivity and can easily be kept and observed in laboratory conditions.

Krebs, J.R. & Davies, N.B. (1981). *An Introduction to Behavioural Ecology.* Oxford: Blackwell Scientific Publications.

A clear and concise introduction to the principal ideas and findings of behavioural ecology.

Lane-Petter, W. (1978). Identification of laboratory animals. In *Animal Marking. Recognition Marking of Animals in Research* (ed. by B. Stonehouse), pp. 35–39. London: Macmillan.

A brief survey of identification methods, including clipping, punching, painting the skin or fur, tattooing, tagging, ringing and the use of collars and belts.

Leger, D.W. (1977). An empirical evaluation of instantaneous and one-zero sampling of chimpanzee behavior. *Primates*, **18**, 387–393.

Recorded the behaviour of chimpanzees using both instantaneous sampling and one-zero sampling (with a 15-s sample interval) and compared the results against a continuous record, for various categories of behaviour. Found that one-zero scores correlated very strongly with corresponding measures of total duration (or percentage time spent performing the behaviour pattern) and moderately with frequency (hourly rate) and mean duration (mean bout length). The correlation with total time decreased with longer sample intervals. One-zero scores were consistently and predictably larger than the actual total duration scores. Instantaneous sampling scores also correlated very strongly ($r = 0.99$) with total duration measures and moderately with frequency and mean duration.

*Lehner, P.N. (1979). *Handbook of Ethological Methods.* New York: Garland STPM.

Deals with most aspects of methodology, from formulating the question to analysing the data. A good general reference book for many topics and well worth browsing through.

Linn, I.J. (1978). Radioactive techniques for small mammal marking. In *Animal Marking. Recognition Marking of Animals in Research* (ed. by B. Stonehouse), pp. 177–191. London: Macmillan.

Discusses the possibilities and hazards of using radioactive markers for tracking animals.

Lockard, R.B. (1971). Reflections on the fall of comparative psychology. *American Psychologist*, **26**, 168–179.
Explains how and why old-style 'comparative' psychology has given way to the ideas of modern behavioural biology, including its attention to functional and evolutionary ideas.

Lott, D.F. (1975). Protestations of a field person. *BioScience*, **25**, 328.
An amusing one-page account of some of the problems faced by field workers.

Lykken, D.T. (1968). Statistical significance in psychological research. *Psychological Bulletin*, **70**, 151–159.
Puts forward the following argument. Many theories about behaviour predict no more than the direction of an effect (such as a difference or correlation). Since the null hypothesis is very rarely true, such predictions have a roughly 50% probability of being confirmed by experiment when the theory is false. The statistical significance of the effect is a function of sample size. Confirmation of a simple directional prediction should therefore add little to the confidence in a theory. Rather, theories should be tested by multiple corroboration. Empirical results should be tested by *constructive replication* – which means using various experimental procedures to produce the same overall empirical conclusions, rather than attempting to replicate the original experiment exactly (*literal replication*). Statistical significance is a relatively minor attribute of a good experiment and is not a sufficient condition for claiming that a theory has been corroborated.

McFarland, D. (1981). Rhythms. In *The Oxford Companion to Animal Behaviour* (ed. by D. McFarland), pp. 478–483. Oxford: Oxford University Press.
A brief survey of rhythms in behaviour, including short-term (ultradian), tidal, circadian, lunar and circannual rhythms.

McFarland, D. (1985). *Animal Behaviour. Psychobiology, Ethology and Evolution*. London: Pitman.
An excellent, clearly written and comprehensive general text on animal behaviour which successfully integrates the ethological and psychological approaches. See especially ch.1 (an introduction to the study of animal behaviour), ch. 6 (which includes a discussion of rhythms in behaviour) and section 3.3 (which deals with animal language, tool use, culture, intelligence, awareness, emotion and various other fascinating issues that would be overlooked by a strict adherent to Lloyd Morgan's Canon).

†McGill, T.E. (ed.) (1977). *Readings in Animal Behaviour*, 3rd edition. New York: Holt, Rinehart & Winston.
A collection of reprints of some important papers from various areas of ethology and comparative psychology.

Macdonald, D.W. (1978). Radio-tracking: some applications and limitations. In *Animal Marking. Recognition Marking of Animals in Research* (ed. by B. Stonehouse), pp. 192–204. London: Macmillan.
A useful discussion of radio-tracking methods.

†Macdonald, D. [W.] & Amlaner, C. (1984). Biotelemetry: listening in to wildlife. In *The Understanding of Animals* (ed. by G. Ferry), pp. 75–86. Oxford: Blackwell/New Scientist.
A popular account of radio tracking and telemetry.

†Macdonald, D.W., Ball, F.G. & Hough, N.G. (1980). The evaluation of home range size and configuration using radio tracking data. In *A Handbook on Biotelemetry and Radio Tracking* (ed. by C.J. Amlaner & D.W. Macdonald), pp. 405–424. Oxford: Pergamon Press.
Uses specimen data to compare conclusions about home range size based on different methods of analysing movement data. Shows that radio tracking can greatly increase the amount of information obtained, but stresses that there is no substitute for field observations.

Machlis, L. (1977). An analysis of the temporal patterning of pecking in chicks. *Behaviour*, **63**, 1–70.
Deals with bouts and log survivorship analysis.

*Machlis, L., Dodd, P.W.D. & Fentress, J.C. (1985). The pooling fallacy: problems arising when individuals contribute more than one observation to the data set. *Zeitschrift für Tierpsychologie*, **68**, 201–214.
Attacks the common practice of obtaining several measurements from each subject and pooling these for statistical analysis. Argues that this procedure is logically flawed and uses simulation to show that pooled data sets are likely to give rise to spuriously significant effects (i.e., the probability of rejecting a true null hypothesis can be substantially greater than the stated level of significance).

Maxwell, A.E. (1977). *Multivariate Analysis in Behavioural Research*. London: Chapman & Hall.
Deals with factor analysis, principal component analysis, multiple regression and other techniques for simultaneously analysing several variables.

Mech, L.D. (1983). *Handbook of Animal Radio-Tracking*. Minneapolis: University of Minnesota Press.
A concise handbook dealing with the theory and practice of radio-tracking. Gives valuable practical information on issues such as the design, range and reliability of transmitters. Includes an extensive bibliography. Essential reading for beginners in radio-tracking.

*Medawar, P.B. (1981). *Advice to a Young Scientist*. London: Pan Books/New York: Harper & Row.

A delightful gold-mine of common sense that should be read by all scientists, regardless of age; short, highly readable and strongly recommended for a spare afternoon.

Meddis, R. (1984). *Statistics Using Ranks. A Unified Approach*. Oxford: Basil Blackwell.
Presents a wide range of nonparametric statistical tests within a unified framework, emphasising the underlying similarity between many tests. Includes a BASIC listing of a computer program that performs most types of nonparametric analysis.

†Michener, G.R. (1980). The measurement and interpretation of interaction rates: an example with adult Richardson's ground squirrels. *Biology of Behaviour*, **5**, 371–384.
Illustrates two problems with quantifying the frequency of social interactions in terms of the number of interactions per individual per unit observation time. First, interactions typically involve two individuals, so the relevant statistic is the number of pairs, not the number of individuals. Secondly, individuals are available to interact only when they are active, so the relevant time unit is the amount of time that both members of a pair are simultaneously active, not the amount of time the observer spends watching.

Morgan, B.J.T., Simpson, M.J.A., Hanby, J.P. & Hall-Craggs, J. (1976). Visualizing interaction and sequential data in animal behaviour: theory and application of cluster-analysis methods. *Behaviour*, **56**, 1–43.
Considers how to condense behavioural data contained in 2-way tables (such as individuals × behaviour scores, or individuals × individuals association measures) using cluster analysis to summarise similarities. Cluster analysis generally makes fewer assumptions than factor analysis or principal component analysis and is easier to understand.

Newby, J.C. (1980). *Mathematics for the Biological Sciences*. Oxford: Clarendon Press.
A book of basic maths for biologists. Contains some short computer programs.

Pennycuick, C.J. (1978). Identification using natural markings. In *Animal Marking. Recognition Marking of Animals in Research* (ed. by B. Stonehouse), pp. 147–159. London: Macmillan.
Discusses the theoretical and practical problems concerned with identifying individuals using natural markings or variations in size, shape, colour, etc.

Purton, A.C. (1978). Ethological categories of behaviour and some consequences of their conflation. *Animal Behaviour*, **26**, 653–670.
Discusses in philosophical terms some of the problems that arise when

behaviour is variously categorised according to its form, function or causation and when these different conceptual schemes are muddled.

†Ransom, R. & Matela, R.J. (1985). *Computers in Biology. An Introduction*. Milton Keynes: Open University Press.

Regal, P.J. & Connolly, M.S. (1980). Social influences on biological rhythms. *Behaviour*, **72**, 171–198.
Considers social effects (such as social synchronisation and desynchronisation) on circadian, ultradian and infradian rhythms in behaviour.

Rhine, R.J. & Flanigon, M. (1978). An empirical comparison of one-zero, focal animal, and instantaneous methods of sampling spontaneous primate social behavior. *Primates*, **19**, 353–361.
Argues that Altmann's (1974) severe criticisms of one-zero sampling are based on theoretical assumptions which are sometimes unnecessarily restrictive or incorrect and are unsupported empirically. Defends the use of one-zero sampling as a valuable and practicable method. An empirical evaluation of one-zero sampling showed that it is a convenient way of combining information about both rate and duration into a single measure. Scores obtained using one-zero sampling and continuous recording were highly correlated. Instantaneous sampling was found to give a very accurate estimate of the proportion of time spent performing an activity.

Rhine, R.J. & Linville, A.K. (1980). Properties of one-zero scores in observational studies of primate social behavior: the effect of assumptions on empirical analyses. *Primates*, **21**, 111–122.
Demonstrates empirically that instantaneous sampling and one-zero sampling can both be highly reliable methods for recording behaviour and refutes criticisms of the validity of one-zero sampling. Argues that there is no conclusive reason for considering frequency and duration as the only valid measures of behaviour, against which time sampling methods must necessarily be compared. Empirical evidence shows that frequency and duration measures of the same behaviour are often poorly correlated, while one-zero scores give a weighted combination of both frequency and duration.

Richards, S.M. (1974). The concept of dominance and methods of assessment. *Animal Behaviour*, **22**, 914–930.
An empirical assessment, based on a study of rhesus monkeys, of the concept of dominance as a single index for predicting a wide variety of social interactions. Different ways of assessing dominance – for example, from priority of access to various food sources, agonistic interactions, displays and various gestures of fear or submission – were found to agree with each other, suggesting that dominance is a useful intervening variable.

Roper, T.J. (1984). Response of thirsty rats to absence of water: frustration, disinhibition or compensation? *Animal Behaviour*, **32**, 1225–1235.

An example of how the intensity of a behaviour pattern can be quantified in terms of its 'local rate'. This is a measure borrowed from operant psychology and is simply the total number of component acts divided by the total time allocated to that activity by the subject. For example, the intensity of eating behaviour could be measured by counting the total number of component acts (say, the consumption of individual food pellets) and dividing this by the total time spent eating during the observation session. The more intense the eating behaviour, the more food pellets are consumed per unit time spent eating.

Rosenthal, R. (1976). *Experimenter Effects in Behavioral Research*. New York: Irvington.

A book-length treatment of the unwanted effects the experimenter or observer has on the outcome of a study. Focuses on expectancy effects (obtaining the expected results) as they occur in laboratory settings and in everyday life.

†Rosenthal, R. (1978). Combining results of independent studies. *Psychological Bulletin*, **85**, 185–193.

Briefly describes different methods for combining the probabilities (significance levels) obtained from two or more independent studies to give a single, overall probability.

†Rosenthal, R. (1979). The 'file drawer problem' and tolerance for null results. *Psychological Bulletin*, **86**, 638–641.

Considers the important problem of unpublished experimental results. In any given area, it is impossible to know how many studies have been conducted but never reported (the 'file drawer problem'). Since studies finding non-significant results are generally less likely to be published than those that do find significant effects, the literature contains a potentially biased sample of the studies that have actually been carried out.

Rosenthal, R. & Rubin, D.B. (1978). Interpersonal expectancy effects: the first 345 studies. *The Behavioral & Brain Sciences*, **1**, 377–415.

Reviews one type of experimenter bias in studies of behaviour: the expectancy effect. This refers to the tendency of experimenters to obtain the results they expected to obtain, partly because they unwittingly influenced the outcome of the experiment. Experimental results therefore become self-fulfilling prophecies. The results of 345 experiments investigating expectancy effects are summarised, including some from animal learning experiments. These show that the expectancy effect undoubtedly exists and is not trivial. Indeed, the expectancy effect is sometimes comparable in size to the effect of the experimental manipulation itself. Expectancy effects are likely to be most important in studies of human behaviour.

Ryan, B.F., Joiner, B.L. & Ryan, T.A. Jr (1985). *MINITAB Handbook*, 2nd edition. Boston, Massachusetts: Duxbury Press.
A student handbook for the MINITAB general purpose statistics package. Also offers clear explanations of statistical concepts and tests and is therefore useful as a supplementary statistics text. MINITAB is an interactive software package that is ideal for beginners: it is easy to use, yet sophisticated enough to deal with most types of analysis. MINITAB, unlike many statistics packages, is good at dealing with exploratory data analysis and also includes facilities for analysing data in the form of time series.

*Sackett, G.P. (ed.) (1978). *Observing Behavior*, Vol. II, *Data Collection and Analysis Methods*. Baltimore: University Park Press.
A useful, multi-author volume dealing with theoretical and methodological aspects of using direct observation of behaviour as a research tool. Primarily concerned with the study of mental retardation and child development. However, many of the topics discussed are of broad significance and relevant to studying the behaviour of other species. Covers many of the basic issues, such as defining categories, sampling techniques, statistical analysis, recording methods and reliability.

Sackett, G.P., Ruppenthal, G.C. & Gluck, J. (1978). Introduction: an overview of methodological and statistical problems in observational research. In *Observing Behavior*, Vol II, *Data Collection and Analysis Methods* (ed. by G. P. Sackett), pp. 1–14. Baltimore: University Park Press.
Includes a brief but informative account of the problem of non-independence of behavioural categories (pp. 12–13).

Sackett, G.P., Stephenson, E. & Ruppenthal, G.C. (1973). Digital data acquisition systems for observing behavior in laboratory and field settings. *Behavior Research Methods & Instrumentation*, **5**, 344–348.
Reviews devices available in the early 1970s for the digital coding of behavioural observations. These involved transcribing data from a keyboard onto magnetic tape, for subsequent analysis on a main-frame computer.

Schleidt, W.M., Yakalis, G., Donnelly, M. & McGarry, J. (1984). A proposal for a standard ethogram, exemplified by an ethogram of the bluebreasted quail (*Coturnix chinensis*). *Zeitschrift für Tierpsychologie*, **64**, 193–220.

Proposes a standardised format for describing the discrete, species-specific behaviour patterns that form the behavioural repertoires of different species. See also the commentaries on this paper in *Zeitschrift für Tierpsychologie*, **68**, 335–345 (1985).

Schneirla, T.C. (1950). The relationship between observation and experimentation in the field study of behavior. *Annals of the New York Academy of Sciences*, **51**, 1022–1044.
An important essay which makes a number of fundamental points about studying behaviour. Argues that no sharp line can be drawn between careful field observation and simple experimental procedure. For example, the use of hides or blinds constitute 'controls' in so far as they reduce the disruptive effects of confounding variables introduced by the observer's presence. Emphasises that field observation and laboratory experiments are overlapping and complementary activities.

Scott, D.K. (1978). Identification of individual Bewick's swans by bill patterns. In *Animal Marking. Recognition Marking of Animals in Research* (ed. by B. Stonehouse), pp. 160–168. London: Macmillan.
An impressive example of how an experienced observer can use natural markings to identify a large number of different individuals.

Scott, J.L. (1970). Studies of maternal and pup behaviour in the golden hamster (*Mesocricetus auratus Waterhouse*). Unpublished PhD dissertation, University of Cambridge.

Seyfarth, R.M., Cheney, D.L. & Marler, P. (1980). Vervet monkey alarm calls: semantic communication in a free-ranging primate. *Animal Behaviour*, **28**, 1070–1094.
Describes results of a field study in Kenya showing that wild vervet monkeys give acoustically distinct alarm calls to three different types of predator (leopard, eagle and snake). These calls elicit different, and apparently adaptive, responses in other individuals (climbing into trees, looking in the air and looking down). Provides a good example of field experiments, in which tape-recorded alarm calls were played to wild vervets in the absence of actual predators.

†Sharman, M. & Dunbar, R.I.M. (1982). Observer bias in selection of study group in baboon field studies. *Primates*, **23**, 567–573.
By examining published data from several field studies of baboon behaviour, Sharman & Dunbar found that previous investigators had consistently tended to study groups that were larger than the mean group size for the population as a whole. They also found evidence that group size is related to certain aspects of behaviour, implying that previous results may have been biased.

Short, R. & Horn, J. (1984). Some notes on factor analysis of behavioral data. *Behaviour*, **90**, 203–214.

If used incorrectly (as they often are), multivariate techniques can be very misleading. This paper draws attention to the pitfalls commonly encountered in the use of factor analysis. Factor analysis should not be used unless all the variables are reliable and the sample size is at least three times greater than the number of variables.

*Siegel, S. (1956). *Nonparametric Statistics for the Behavioral Sciences*. New York: McGraw-Hill.
Still the best single text on nonparametric methods; clear, concise, reliable and indispensable.

Simon, T. & Smith, P.K. (1983). The study of play and problem solving in preschool children: have experimenter effects been responsible for previous results? *British Journal of Developmental Psychology*, **1**, 289–297.
Shows how failure to use adequate control groups and prevent observer bias can invalidate experiments.

Simpson, M.J.A. (1979). Daytime rest and activity in socially living rhesus monkey infants. *Animal Behaviour*, **27**, 602–612.
Empirical results from a study of rhesus monkeys showed that the quality as well as quantity of mother–infant contact differed considerably between daytime periods of rest and activity.

Simpson, M.J.A. & Simpson, A.E. (1977). One-zero and scan methods for sampling behaviour. *Animal Behaviour*, **25**, 726–731.
Shows empirically, using data describing mother-infant interactions in rhesus monkeys, that instantaneous sampling can give good estimates of durations, but casts doubt on the suitability of one-zero sampling as a method for estimating total duration. Confusingly refers to instantaneous sampling as 'scan' sampling.

Slater, P.J.B. (1973). Describing sequences of behavior. In *Perspectives in Ethology*, Vol. 1 (ed. by P.P.G. Bateson & P.H. Klopfer), pp. 131–153. New York: Plenum Press.
A clear account of methods for analysing sequences of behaviour, and their limitations.

Slater, P.J.B. (1974). The temporal pattern of feeding in the zebra finch. *Animal Behaviour*, **22**, 506–515.
An example of log survivorship analysis of bout length.

*Slater, P.J.B. (1978). Data collection. In *Quantitative Ethology* (ed. by P.W. Colgan), pp. 7–24. New York: Wiley.
Considers why measurement of behaviour is necessary and discusses categorisation and sampling of behaviour.

Slater, P.J.B. (1980). The relevance of ethology. In *Developmental Psychology and Society* (ed. by J. Sants), pp. 96–119. London: Macmillan.

An essay discussing the relevance of ethological ideas and observational methods to psychology.

Slater, P.J.B. (1981). Individual differences in animal behavior. In *Perspectives in Ethology*, Vol. 4 (ed. by P.P.G. Bateson & P.H. Klopfer), pp. 35–49. New York: Plenum Press.
Considers why individual differences in behaviour may make adaptive sense. For example, differences in feeding, mating or fighting behaviour may have arisen because natural selection has favoured the use of different strategies by different individuals within a species. Individual variation in signals may function to signal the identity of the individual. Some variation, however, may have arisen because the exact form of the behaviour makes little difference in terms of selection or because the optimal behaviour cannot be forecast in an unpredictable environment.

Slater, P.J.B. (1983). The study of communication. In *Animal Behaviour*, Vol. 2, *Communication* (ed. by T.R. Halliday & P.J.B. Slater), pp. 9–42. Oxford: Blackwell Scientific Publications.
Includes a succinct account of analysing sequences of behaviour.

Slater, P.J.B. & Lester, N.P. (1982). Minimising errors in splitting behaviour into bouts. *Behaviour*, **79**, 153–161.
Discusses problems with the conventional method of performing log survivorship analysis to estimate bout length.

†Smith, P.K. (1974). Ethological methods. In *New Perspectives in Child Development* (ed. by B. Foss), pp. 85–137. Harmondsworth, Middlesex: Penguin Education.
Reviews the applications of ethological methods of observation to the study of children's behaviour.

Smith, P.K. (1985). The reliability and validity of one-zero sampling: misconceived criticisms and unacknowledged assumptions. *British Educational Research Journal*, **11**, 215–220.
A compelling defence of one-zero time sampling as a reliable and valid method for recording behaviour.

Smith, P.K. & Connolly, K.J. (1980). *The Ecology of Preschool Behaviour*. Cambridge: Cambridge University Press.
A good example of how ethological methods of observation have been applied successfully to the study of children's behaviour. (Also includes a defence of one-zero time sampling.)

Smith, P.K. & Simon, T. (1984). Object play, problem-solving and creativity in children. In *Play in Animals and Humans* (ed. by P.K. Smith), pp. 199–216. Oxford: Basil Blackwell.
Deals with the general problems of experimenter bias and selecting

the appropriate control groups, in relation to experimental studies of children's play.

Snedecor, G.W. & Cochran, W.G. (1980). *Statistical Methods*, 7th edition. Ames, Iowa: Iowa State University Press.
One of the best general texts on statistics; comprehensive, concise and good for reference.

Sokal, R.R. & Rohlf, F.J. (1981). *Biometry*, 2nd edition. San Francisco: W.H. Freeman.
Another widely used statistics text.

*Sprinthall, R.C. (1982). *Basic Statistical Analysis*. Reading, Massachusetts: Addison-Wesley.
Excellent for providing easily understood explanations of the basic concepts of statistics and experimental design.

Stevenson-Hinde, J. (1983). Individual characteristics and the social situation. In *Primate Social Relationships. An Integrated Approach* (ed. by R.A. Hinde), pp. 28–35. Oxford: Blackwell Scientific Publications.
A brief explanation of methods for assessing personality; reviews work on subjective assessment of individual rhesus monkeys.

Stevenson-Hinde, J., Stillwell-Barnes, R. & Zunz, M. (1980). Subjective assessment of rhesus monkeys over four successive years. *Primates*, **21**, 66–82.
This study demonstrated that observers' ratings of various aspects of individual monkeys' personality can be both reliable and valid. Observers rated the same individual monkeys in four successive years, using behaviourally-defined adjectives and a 7-point rating scale.

Still, A.W. (1982). On the number of subjects used in animal behaviour experiments. *Animal Behaviour*, **30**, 873–880.
Argues that the number of animals used in behavioural experiments could often be reduced without loss of scientific rigour; sample size can sometimes be reduced if experimental design is improved. Discusses the logic of inference from studies with small sample sizes.

Stonehouse, B. (ed.) (1978). *Animal Marking. Recognition Marking of Animals in Research*. London: Macmillan.
A useful reference covering most aspects of marking, including methods of capture, marking in captivity, tagging, recognition by natural markings, radioactive marking, radio tracking and biotelemetry. Deals with applications of these techniques to mammals, birds, reptiles, fish and invertebrates.

Symonds, R.J. & Unwin, D.M. (1982). The use of a microcomputer to collect activity data. *Physiological Entomology*, **7**, 91–98.
Explains in simple terms how a microcomputer can be used to record data from automatic activity detectors.

Ten Cate, C. (1985). Directed song of male zebra finches as a predictor of subsequent intra- and interspecific social behaviour and pair formation. *Behavioural Processes*, **10**, 369–374.
An example of how a behavioural variable can be validated by demonstrating empirically that it does indeed measure the things it is supposed to measure. In studies of sexual imprinting, the amount of song a male bird directs towards a female is generally taken as a measure of sexual preference. Ten Cate has validated this measure by showing that the amount of singing directed to different females does indeed predict subsequent measures of social and sexual behaviour, including pair formation.

*Tinbergen, N. (1963). On aims and methods of ethology. *Zeitschrift für Tierpsychologie*, **20**, 410–433.
Words of wisdom from one of the founders of ethology and essential reading for any aspiring ethologist. Outlines the distinction between the 'Four Problems' (proximate causation, function, evolution and development) and considers many basic issues in ethology.

Tinbergen, N., Broekhuysen, G.J., Feekes, F., Houghton, J.C.W., Kruuk, H. & Szulc, E. (1962). Egg shell removal by the black-headed gull, *Larus ridibundus* L.; a behaviour component of camouflage. *Behaviour*, **19**, 74–117.
The classic example of a field experiment designed to test a functional hypothesis about a behaviour pattern. A summary of this work is given in 'The Shell Menace', by N. Tinbergen (1963), *Natural History*, **72**, 28–35, which is reprinted in the book of readings edited by McGill (1977).

Tobach, E., Schneirla, T.C., Aronson, L.R. & Laupheimer, R. (1962). The ATSL: an observer-to-computer system for a multivariate approach to behavioural study. *Nature (London)*, **194**, 257–258.
An early example of a behavioural event recorder. The keyboard consisted of a bank of microswitches, each coding for one behavioural measure, and enabling the observer to record up to 40 different measures. Data were coded onto paper punch-tape, for subsequent analysis on a computer.

Trivers, R. (1985). *Social Evolution*. Menlo Park, California: Benjamin/Cummings.

Tukey, J.W. (1977). *Exploratory Data Analysis*. Reading, Massachusetts: Addison-Wesley.
The principal text on exploratory data analysis. Unfortunately, the idiosyncratic style of presentation makes it difficult reading.

Twigg, G.I. (1978). Marking mammals by tissue removal. In *Animal Marking. Recognition Marking of Animals in*

Research (ed. by B. Stonehouse), pp. 109–118. London: Macmillan.

Reviews marking by ear-clipping, toe-clipping, fur-clipping, tail-docking, heat branding, freeze branding and tattooing.

*Tyler, S. (1979). Time-sampling: a matter of convention. *Animal Behaviour*, **27**, 801–810.

A sensible account of time-sampling methods. Shows empirically that both one-zero and instantaneous sampling can be either accurate or inaccurate, depending on what is being measured, and points out the many practical advantages of time sampling.

UFAW (ed.) (1976). *The UFAW Handbook on the Care and Management of Laboratory Animals*, 5th edition. Edinburgh: Churchill Livingstone.

A valuable reference covering the natural history, care and housing of many laboratory species.

Vauclair, J. & Bateson, P.P.G. (1975). Prior exposure to light and pecking accuracy in chicks. *Behaviour*, **52**, 196–201.

Velleman, P.F. & Hoaglin, D.C. (1981). *Applications, Basics, and Computing of Exploratory Data Analysis*. Boston, Massachusetts: Duxbury Press.

An introductory text on EDA techniques such as stem-and-leaf displays, boxplots, smoothing data and computer graphics; includes BASIC and FORTRAN listings of computer programs for the principal techniques.

White, L.E. (1977). The nature of social play and its role in the development of the rhesus monkey. Unpublished PhD dissertation, University of Cambridge.

White, R.E.C. (1971). WRATS: A computer compatible system for automatically recording and transcribing behavioural data. *Behaviour*, **40**, 135–161.

An early example of a sophisticated, custom-built event-recording system. Observations were recorded on a keyboard with up to 40 keys and stored on magnetic tape prior to analysis on a computer. The system had a time resolution of 0.1 s.

†Wiley, R.H. (1973). The strut display of male sage grouse: a 'fixed' action pattern. *Behaviour*, **47**, 129–152.

The strut display of the male sage grouse is one of the few well-documented examples of a species-specific, stereotyped behaviour pattern that is highly invariant in form.

Wohlwill, J.F. (1973). *The Study of Behavioral Development*. New York: Academic Press.

A book-length treatment of the theoretical issues underlying methods for studying behavioural development. Aimed mainly at developmental psychologists, but also relevant to those studying development in non-human species. Includes chapters on problems of measurement in developmental psychology, longitudinal versus cross-sectional methodology, correlational methods in the study of developmental change, the experimental manipulation of developmental change and individual differences in development.

Wood-Gush, D.G.M. (1983). *Elements of Ethology. A Textbook for Agricultural and Veterinary Students*. London: Chapman & Hall.
A basic, introductory text for agricultural and veterinary students; ch. 14 covers animal welfare.

Zar, J.H. (1984). *Biostatistical Analysis*, 2nd edition. New York: Prentice-Hall.
One of the best general texts on statistics.

INDEX

Where appropriate, primary entries are indicated in bold.